Conference Proceedings of the Society for Experimental Mechanics Series

Series Editor
Kristin B. Zimmerman
Society for Experimental Mechanics, Inc.,
Bethel, CT, USA

The Conference Proceedings of the Society for Experimental Mechanics Series presents early findings and case studies from a wide range of fundamental and applied work across the broad range of fields that comprise Experimental Mechanics. Series volumes follow the principle tracks or focus topics featured in each of the Society's two annual conferences: IMAC, A Conference and Exposition on Structural Dynamics, and the Society's Annual Conference & Exposition and will address critical areas of interest to researchers and design engineers working in all areas of Structural Dynamics, Solid Mechanics and Materials Research.

Matthew Allen • Walter D'Ambrogio • Dan Roettgen
Editors

Dynamic Substructures, Volume 4

Proceedings of the 40th IMAC, A Conference and Exposition on Structural Dynamics 2022

Springer

Editors
Matthew Allen
Brigham Young University
Provo, UT, USA

Dan Roettgen
Sandia National Laboratories
Albuquerque, NM, USA

Walter D'Ambrogio
DIIIE
University of L'Aquila
L'AQUILA, L'Aquila, Italy

ISSN 2191-5644 ISSN 2191-5652 (electronic)
Conference Proceedings of the Society for Experimental Mechanics Series
ISBN 978-3-031-04093-1 ISBN 978-3-031-04094-8 (eBook)
https://doi.org/10.1007/978-3-031-04094-8

This Springer imprint is published by the registered company Springer Nature Switzerland AG
The registered company address is: Gewerbestrasse 11, 6330 Cham, Switzerland

Preface

Dynamic Substructures represents one of nine volumes of technical papers presented at the 40th IMAC, A Conference and Exposition on Structural Dynamics, organized by the Society for Experimental Mechanics, and held February 7–10, 2022. The full proceedings also include volumes titled *Nonlinear Structures & Systems*; *Dynamics of Civil Structures*; *Model Validation and Uncertainty Quantification*; *Special Topics in Structural Dynamics & Experimental Techniques*; *Rotating Machinery, Optical Methods & Scanning LDV Methods*; *Sensors and Instrumentation, Aircraft/Aerospace and Dynamic Environments Testing*; *Topics in Modal Analysis & Parameter Identification*; *and Data Science in Engineering*.

Each collection presents early findings from experimental and computational investigations on an important area within structural dynamics. Coupled structures or, substructuring, is one of these areas.

Substructuring is a general paradigm in engineering dynamics where a complicated system is analyzed by considering the dynamic interactions between subcomponents. In numerical simulations, substructuring allows one to reduce the complexity of parts of the system in order to construct a computationally efficient model of the assembled system. A subcomponent model can also be derived experimentally, allowing one to predict the dynamic behavior of an assembly by combining experimentally and/or analytically derived models. This can be advantageous for subcomponents that are expensive or difficult to model analytically. Substructuring can also be used to couple numerical simulation with real-time testing of components. Such approaches are known as hardware-in-the-loop or hybrid testing.

Whether experimental or numerical, all substructuring approaches have a common basis, namely the equilibrium of the substructures under the action of the applied and interface forces and the compatibility of displacements at the interfaces of the subcomponents. Experimental substructuring requires special care in the way the measurements are obtained and processed in order to assure that measurement inaccuracies and noise do not invalidate the results. In numerical approaches, the fundamental quest is the efficient computation of reduced order models describing the substructure's dynamic motion. For hardware-in-the-loop applications, difficulties include the fast computation of the numerical components and the proper sensing and actuation of the hardware component. Recent advances in experimental techniques, sensor/actuator technologies, novel numerical methods, and parallel computing have rekindled interest in substructuring in recent years leading to new insights and improved experimental and analytical techniques.

The organizers would like to thank the authors, presenters, session organizers, and session chairs for their participation in this track.

Provo, UT, USA — Matthew Allen
L'Aquila, Italy — Walter D'Ambrogio
Albuquerque, NM, USA — Dan Roettgen

Contents

1 **Uncertainty in Power Flow Due to Measurement Errors in Virtual Point Transformation for Frequency-Based Substructuring** 1
Jon Young and Kyle Myers

2 **Using Flight Test Measurements on a Low-Fidelity Component to Predict Response on a High-Fidelity Component** 11
Randall L. Mayes

3 **Road Noise NVH Part 2: Exploring the Capabilities of the TPA Framework with Interface Forces** 19
Julie M. Harvie, Maarten V. van der Seijs, David P. Song, and Munhwan Cho

4 **Real-Time Hybrid Substructuring for Shock Applications Considering Effective Actuator Control** 29
Christina Insam, Michael J. Harris, Matthew R. Stevens, and Richard E. Christenson

5 **Hrishikesh S. Gosavi, Phanisri P. Pratapa, and Vijaya V. N. Sriram Malladi** 41
Hrishikesh S. Gosavi, Phanisri P. Pratapa, and Vijaya V. N. Sriram Malladi

6 **An Assessment on the Efficiency of Different Reduction Techniques Based on Substructuring for Bladed Disk Systems with Shrouds** 49
Ehsan Naghizadeh and Ender Cigeroglu

7 **Sandor Beregi, David A. W. Barton, Djamel Rezgui, and Simon A. Neild** 59
Sandor Beregi, David A. W. Barton, Djamel Rezgui, and Simon A. Neild

8 **Accuracy of Nonlinear Substructuring Technique in the Modal Domain** 63
Jacopo Brunetti, Walter D'Ambrogio, Annalisa Fregolent, and Francesco Latini

9 **Quantification of Bias Errors Influence in Frequency Based Substructuring Using Sensitivity Analysis** 71
Gregor Čepon, Domen Ocepek, Jure Korbar, Tomaž Bregar, and Miha Boltežar

10 **Real-Time and Pseudo-Dynamic Hybrid Simulation Methods: A Tutorial** 75
Oh-Sung Kwon and Vasilis Dertimanis

11 **Identification of Bolted Joint Properties Through Substructure Decoupling** 85
Jacopo Brunetti, Walter D'Ambrogio, Matteo Di Manno, Annalisa Fregolent, and Francesco Latini

12 **Parametric Analysis of the Expansion Process Based on System Equivalent Model Mixing** 97
Miha Kodrič, Tomaž Bregar, Gregor Čepon, and Miha Boltežar

13 **Real-Time Hybrid Simulation Study of a Physical Duffing Absorber Attached to a Virtual Nonlinear Structure** 101
A. Mario Puhwein and Markus J. Hochrainer

14 System Equivalent Model Mixing (SEMM): A Modal Domain Formulation 111
Miha Pogačar, Domen Ocepek, Gregor Čepon, and Miha Boltežar

15 Hybrid Testing of a Cantilever Beam with Two Controlled Degrees of Freedom 115
Alessandra Vizzaccaro, Sandor Beregi, David Barton, and Simon Neild

16 Experimental Substructuring of the Dynamic Substructures Round-Robin Testbed 119
D. Roettgen, G. Lopp, A. Jaramillo, and B. Moldenhauer

17 Feasibility of Configuration-Dependent Substructure Decoupling 125
Jacopo Brunetti, Walter D'Ambrogio, and Annalisa Fregolent

Chapter 1
Uncertainty in Power Flow Due to Measurement Errors in Virtual Point Transformation for Frequency-Based Substructuring

Jon Young and Kyle Myers

Abstract Experimental substructuring by means of virtual point transformation (VPT) can be utilized to implicitly account for rotational degrees of freedom (DOFs) at the coupling boundary of a given substructure. Measured frequency response functions (FRFs) are cast onto a "virtual" node containing three translational and three rotational DOFs via a projection matrix, which is determined by geometric relations between impact and response locations on the structure. In this study, the uncertainty of power flow due to measurement errors in the projection matrix and FRFs is quantified by means of Monte-Carlo simulation. This is performed on a numerical model of two beam structures and is then compared to experimentally obtained data using an impact modal test. Upper and lower bounds on the broadband power flow are created using these data. It is shown that closely space modes lead to high variability in the calculation of power flow near resonance even for small measurement errors. Metrics for analyzing the quality of the virtual point transformation are discussed, as well. This work is beneficial to understanding how experimental errors manifest in the calculation of power flow between coupled structures.

Keywords Power flow · Dynamic substructuring · Virtual point transformation · Monte-Carlo · Uncertainty quantification

1.1 Introduction

Frequency-based substructuring has been thoroughly developed over the past several decades for the steady-state analysis of numerical structural models. Its application to experimental data, however, leads to several difficulties which, in general, make experimental substructuring a more cumbersome means of predicting the dynamics of complex structures [1]. A set of frequency response functions (FRFs) are used to define the dynamic properties of the structures of interest when they are uncoupled from each other [2–4]. These FRFs are then assembled by enforcing displacement compatibility and force equilibrium on their coupling interfaces, and the resulting FRFs predict the frequency response of the assembled structures. These FRFs are generally obtained using modal impact testing, but shakers or laser vibrometers can also be used [5]. Of course, measurement errors due to noise and uncertainty in the exact position and orientation of accelerometers and impacts are inevitable, but the measurements of rotational degrees of freedom (DOFs) which contain necessary information for proper substructure coupling are generally impossible to measure. Moreover, these errors are drastically amplified during substructure assembly due to the inversion of the interface FRFs and further magnified if power flow between substructures is calculated due to its quadratic form.

To account for the rotational DOFs on the substructure interfaces, the virtual point transformation (VPT) has been developed and successfully used to couple experimental substructures [6–8]. It implements a least-squares residual minimization to project the dynamics measured by accelerometers onto a virtual point which contains three translational and three rotational DOFs. The projection is done using only the geometric relations between the virtual point and the interface impact and response measurements. The primary assumption is that the interface being modelled with a virtual point behaves rigidly in the frequency range being analyzed. That is, there is no local flexible deformation of the interface.

J. Young (✉)
Department of Mechanical Engineering, Pennsylvania State University, State College, PA, USA

K. Myers
Structural Acoustics Department, Penn State Applied Research Laboratory, State College, PA, USA
e-mail: krm25@arl.psu.edu

M. Allen et al. (eds.), *Dynamic Substructures, Volume 4*, Conference Proceedings of the Society for Experimental Mechanics Series,
https://doi.org/10.1007/978-3-031-04094-8_1

Metrics such as overall sensor and impact consistency have been developed to quantify this assumption for a given virtual point [9].

The VPT has been developed to be applied to the Lagrange Multiplier Frequency-Based Substructuring (LMFBS) method [3]. It is known as a dual substructure assembly method because in addition to the displacements of the assembled structure being solved for, the coupling force between substructure interfaces is explicitly solved for. This is particularly useful when calculating power flow because it can be expressed in terms of the coupling interface mobility and the coupling force between substructures. Power flow between substructures through flexible joints has been studied in the past using the LMFBS as the method of assembly [10]. However, this study focused on correcting a numerical model with a flexible elastic and dissipative interface so the response of the assembly would better match experimental data. This type of interface modelling will not be used in this work, and instead a rigid connection is assumed.

Since the VPT requires the projection of physical measurements onto a virtual node near the coupling interface of a substructure, it would be beneficial to know how measurement errors in this projection propagate through the calculation of a response quantity. In this study, power flow from a source to receiver beam structure is the response quantity of interest because it is a single scalar measure of the response of the assembled structures. Measurement uncertainty in both the projection matrix and FRFs is tracked through the substructure assembly process and calculation of power flow, and distributions of power flow and overall sensor consistency are estimated using the results from a Monte-Carlo simulation. Additionally, different ranges of maximum uncertainty in the projection matrix measurements are studied to understand how variability in power flow changes with respect to the amount of uncertainty in the VPT. These simulated results are then compared to some preliminary experimental data.

1.2 Virtual Point Transformation

The VPT projects measured displacements and forces onto a virtual point with three translational and three rotational DOFs. The projection of the response measured by accelerometer k onto virtual point v is given by

$$\mathbf{u}^k = \mathbf{R}_{\mathbf{u}}^{kv}\mathbf{q}^v + \boldsymbol{\mu}^k \tag{1.1}$$

where

$$\mathbf{R}_{\mathbf{u}}^{kv} = \begin{bmatrix} e^k_{x,X} & e^k_{x,Y} & e^k_{x,Z} \\ e^k_{y,X} & e^k_{y,Y} & e^k_{y,Z} \\ e^k_{z,X} & e^k_{z,Y} & e^k_{z,Z} \end{bmatrix} \begin{bmatrix} 1\,0\,0 & 0 & r^k_Z & -r^k_Y \\ 0\,1\,0 & -r^k_Z & 0 & r^k_X \\ 0\,0\,1 & r^k_Y & -r^k_X & 0 \end{bmatrix} \tag{1.2}$$

is the displacement projection matrix. In previous literature, this matrix is said to contain the *interface deformation modes*. The vector $\mathbf{u}^k$ are the physical displacements measured by accelerometer k, and the vector $\mathbf{q}^v$ are the virtual point displacements of virtual point v. This vector contains three translational and three rotational DOFs. Since the VPT assumes a locally rigid interface, the vector μ^k contains the residual flexible motion of the interface measured by accelerometer k. The left matrix in Eq. 1.2 corresponds to the orientation of the accelerometer in the virtual point coordinate system, and the right matrix contains the projection information for the translational and rotational DOFs of the virtual point. A similar projection is performed for the interface forces applied to the structure

$$\mathbf{m}^v = \mathbf{R}_{\mathbf{f}}^{vh} \left\|\mathbf{f}^h\right\| \tag{1.3}$$

where

$$\mathbf{R}_{\mathbf{f}}^{hv} = \left\{ e^h_X \; e^h_Y \; e^h_Z \right\} \begin{bmatrix} 1\,0\,0 & 0 & r^h_Z & -r^h_Y \\ 0\,1\,0 & -r^h_Z & 0 & r^h_X \\ 0\,0\,1 & r^h_Y & -r^h_X & 0 \end{bmatrix} \tag{1.4}$$

is the force projection matrix. Note that the transpose of the force projection matrix is given in Eq. 1.3 due to the virtual point moments not being able to be uniquely defined given only impact force magnitude $\|\mathbf{f}^h\|$ and orientation. Here, $\mathbf{m}^v$ are the virtual forces on virtual point v, and $\mathbf{f}^h$ is the force vector of interface impact h whose direction is given by $\left\{ e^h_X \; e^h_Y \; e^h_Z \right\}$.

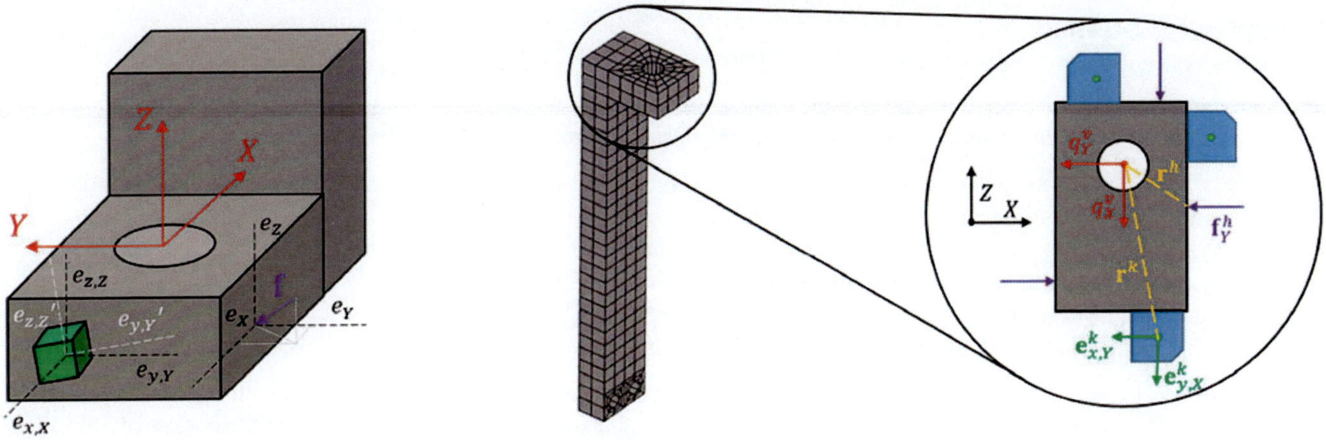

Fig. 1.1 Perturbations in FRFs considered with accelerometer in green and applied interface impact in purple (left); VPT coordinate system and quantities needed to construct projection matrices (right)

In Eq. 1.2, $e_{c,D}^{k}$ and r_D^k are the local direction and distance vectors for accelerometer k, respectively. That is, $e_{c,D}^{k}$ is the local direction c of accelerometer k expressed in the virtual point coordinate direction D of virtual point v. Additionally, in Eq. 1.4, e_D^h and r_D^h are the direction and distance vectors for applied force h. Since impact forces are only measured in a single direction, as compared to a tri-axial accelerometer which measures the response in three orthogonal directions, there is no local coordinate system associated with the applied force. Its direction is simply expressed in the virtual point coordinate system.

Geometric relations between both the interface displacements and forces are given in Fig. 1.1 (right). The VPT is performed by determining a transformation matrix which relates the physical and virtual coordinates. This is done in a least-squares sense using the pseudo-inverse giving

$$\mathbf{q} = \left(\mathbf{R}_{\mathbf{u}}^T\mathbf{R}_{\mathbf{u}}\right)^{-1}\mathbf{R}_{\mathbf{u}}^T\mathbf{u} = \mathbf{T}_{\mathbf{u}}\mathbf{u} \tag{1.5}$$

$$\mathbf{f} = \left(\mathbf{R}_{\mathbf{f}}\mathbf{R}_{\mathbf{f}}^T\right)^{-1}\mathbf{R}_{\mathbf{f}}\mathbf{m} = \mathbf{T}_{\mathbf{f}}^T\mathbf{m} \tag{1.6}$$

These transformations can then be applied to the LMFBS to calculate the virtual point displacements. Doing so gives

$$\mathbf{q}_{\text{ass}} = \mathbf{T}_{\mathbf{u}}\mathbf{Y}\mathbf{T}_{\mathbf{f}}^T\left(\mathbf{f} - \mathbf{B}_{\mathbf{f}}^T\boldsymbol{\lambda}\right) \tag{1.7}$$

where $\mathbf{q}_{\text{ass}}$ contains the virtual point displacements of the assembled structures and

$$\boldsymbol{\lambda} = \left(\mathbf{B}_{\mathbf{u}}\mathbf{T}_{\mathbf{u}}\mathbf{Y}\mathbf{T}_{\mathbf{f}}^T\mathbf{B}_{\mathbf{f}}^T\right)^{-1}\mathbf{B}_{\mathbf{u}}\mathbf{T}_{\mathbf{u}}\mathbf{Y}\mathbf{f} \tag{1.8}$$

is the coupling force between the virtual point connections. This quantity is used to calculate power flow. The matrices $\mathbf{B}_{\mathbf{u}}$ and $\mathbf{B}_{\mathbf{f}}$ identify the virtual point partition of the FRFs in the event that internal DOFs on either the source or receiver are retained, and $\mathbf{Y}$ is a block diagonal matrix containing the source and receiver FRFs. Power can then be calculated using

$$P = \tfrac{1}{2}\,\text{Re}\left\{\boldsymbol{\lambda}^H\mathbf{Y}_{vv}^{(r)}\boldsymbol{\lambda}\right\} \tag{1.9}$$

where the subscript vv indicates the virtual point FRF partition and the superscript (r) indicates a quantity pertaining to the receiver beam. The response of this quantity to measurement uncertainty in the projection matrices is what is to be studied via Monte-Carlo simulation.

1.3 Numerical Methods

A sampling size of 100,000 power flow calculations was used in the Monte-Carlo simulation over a frequency spectrum ranging from 60 to 6000 Hz, increasing in increments of 1 Hz. Four different test cases were simulated:

1. Perturb only r_D^k and r_D^h of the projection matrices, with maximum measurement uncertainty (error) ranging from $\pm 0.05''$ to $\pm 0.45''$, increasing in increments of $0.1''$.
2. Perturb only the values of $e_{c,D}^k$, e_D^h, r_D^k, and r_D^h of the projection matrices, whose ranges are specified in Table 1.1.
3. Perturb only the angles of the impact forces and accelerometers on the interface (effectively perturbing the FRFs) whose ranges are specified in Table 1.1.
4. Perturb both the projection matrices and the FRFs in the same manner specified in the previous two cases.

Case 3 is shown in Fig. 1.1 (left). The 100,000 power flows calculated at a given frequency was allocated to 1 of the 120 bins ranging from the maximum to minimum power flow calculated at that frequency. The distribution of overall sensor consistency was determined in this way as well. Overall sensor consistency was calculated by applying a force to the uncoupled source and receiver sufficiently far from the interface and averaging the source and receiver consistencies as

$$\rho_{\text{avg}} = \frac{1}{2}\left(\rho^{(s)} + \rho^{(r)}\right) \tag{1.10}$$

where the overall sensor consistency of interface displacements is given by

$$\rho = \frac{\|\mathbf{R_u q}\|}{\|\mathbf{u}\|} \tag{1.11}$$

which is a measure of how rigid the interface is and therefore how accurate the VPT is in representing the interface dynamics. It is dimensionless and bounded between 0 and 1. Flexible interfaces will have a sensor consistency close to 0, and rigid interfaces will have a consistency close to 1. Perfectly rigid interfaces have a consistency of exactly 1.

In order to determine if the VPT applied to the numerical model accurately represented the dynamics of the assembled source and receiver beams in a numerical setting, a modal analysis of the assembled finite element models was performed in NASTRAN. The two beams were connected using a rigid body element (RBE2 in NASTRAN) to simulate the virtual point connection. It was verified that the first 16 natural frequencies of the assembled structures using the VPT were within 0.025% of the natural frequencies predicted by coupling the structures with the RBE2 (Fig. 1.2).

Due to the measurement uncertainties simulated, it is possible to calculate negative power flows using Eq. 1.9. However, values of power flow below −240 dB were truncated from the data set as negative values cannot be shown on a logarithmic

Table 1.1 Maximum and minimum uncertainties in projection matrix and FRFs for Cases 2–4 of Monte-Carlo simulation

	Projection Matrix				FRFs	
	Accel.		Force			
	Location (in)	Angle (deg)	Location (in)	Angle (deg)	Accel. Angle (deg)	Force. Angle (deg)
Min	−0.25	−2	−0.25	−2	−2	−10
Max	0.25	2	0.25	2	2	10

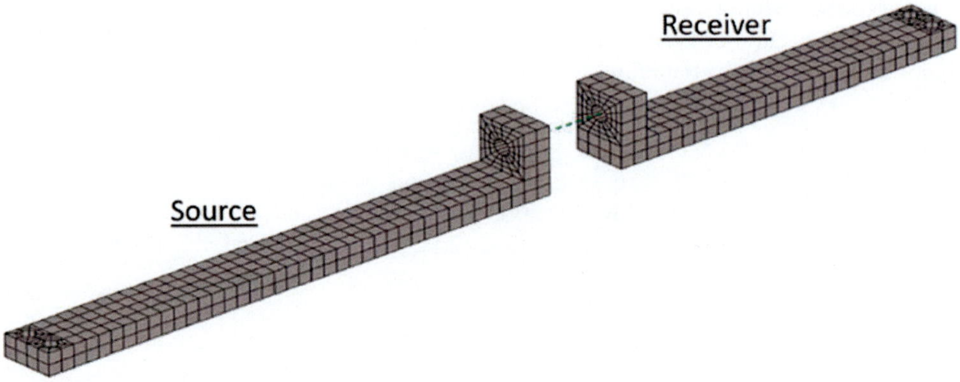

Fig. 1.2 Numerical model of source and receiver beam structures to be coupled using VPT using a single virtual point

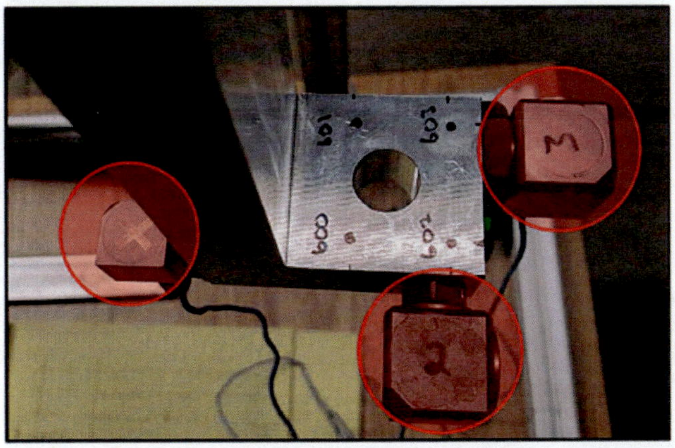

Fig. 1.3 Experimental interface measurements showing three tri-axial accelerometers (highlighted in red) and several impact locations indicated by the points labeled 600–603

scale. Negative power flows will be discussed with the results from Monte-Carlo simulation and the calculation of power flow using experimental data.

1.4 Experimental Methods

Power flow was estimated using experimental data by measuring the response of the uncoupled source and receiver interfaces using three tri-axial accelerometers (two PCB-B07 and one PCB-A14 due to equipment limitations), shown in Fig. 1.3. A total of 15 impacts were made to each interface which were to be projected onto the virtual point. An impact modal hammer (model PCB-086C01) with a metal tip was used to measure the force applied to the interface. A sampling rate of 1 Hz was used at frequencies ranging from 60 to 1600 Hz. This range was sufficient to capture five modes of the assembled substructures predicted by the numerical model, excluding rigid body modes.

Due to their size, it should be mentioned that the mass loading of accelerometers was included in the numerical model described in the previous section. The two PCB-B07 accelerometers weighed approximately 35 g each, and the one PCB-A14 accelerometer weighed 12 g. It will be seen in the results that the mass of the accelerometers on the interface played a much more prominent role in predicting the assembled response of the structure than could be modelled by simply attaching point masses to the interface of the numerical model.

1.5 Results

The effect of measurement uncertainty in the distance parameters in the VPT is shown in Fig. 1.4 at several frequencies. Practically speaking, large variability in power flow can occur due to relatively small errors present when measuring the distance between the virtual point, force, and accelerometer locations in an experimental setup. The distribution of power flow for different ranges of measurement uncertainty is shown at 850 Hz in Fig. 1.4 (left) – effectively Case 1 of the simulation. It can be seen that the statistical mean, media, mode, and standard deviation of power flow change as a function of the uncertainty in the distance parameters of the projection matrices. Additionally, the distribution of power flow in Fig. 1.4 is not Gaussian. The shape of the distribution was heavily dependent on frequency, and while at some frequencies it was heavily skewed to the left or right, at other frequencies, it was approximately Gaussian. Large changes in the distribution of power flow occurred near resonance, but a description of the distribution as a function of frequency has not been developed.

The standard deviation of the power flow distributions is shown at several frequencies in Fig. 1.4 (right). The first frequency (475 Hz) coincides with the second resonant frequency of the assembled system, the second frequency (850 Hz) is between the third and fourth resonant frequencies, the third frequency (2.6 kHz) is near two closely spaced modes, and the fourth frequency (5.5 kHz) is in a frequency region with decreased sensor consistency. The first, second, and fourth frequencies are directly proportional to the measurement uncertainty in the projection matrix distance parameters. The third frequency, near two closely spaced modes, has the highest standard deviation when the maximum measurement uncertainty

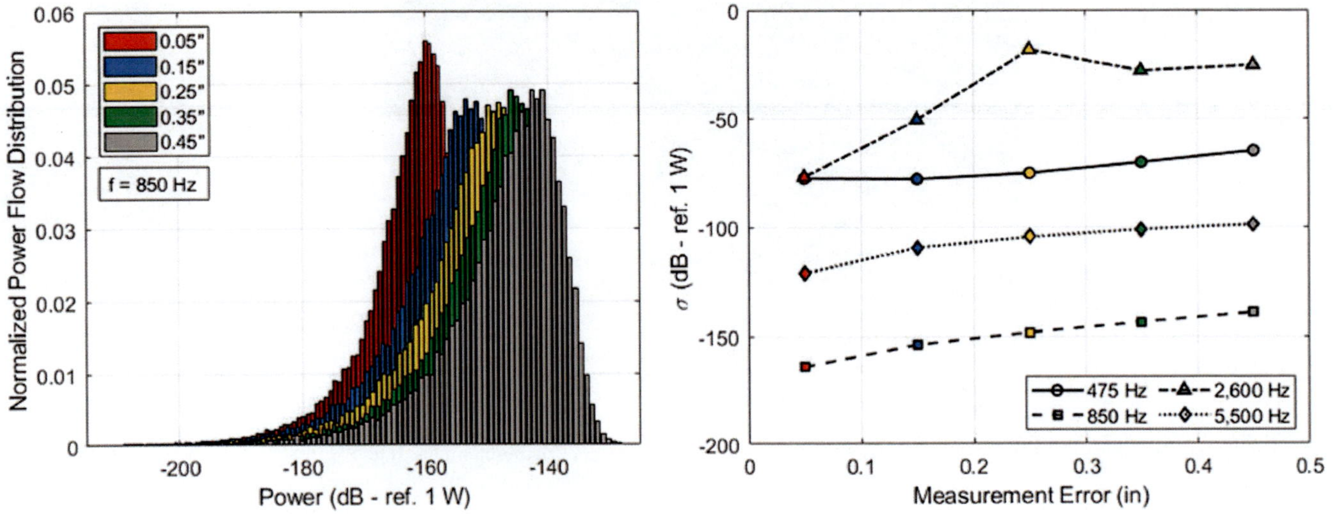

Fig. 1.4 Distribution of power flow normalized by maximum value of distribution across entire frequency spectrum for different ranges of measurement error at 850 Hz (left), Standard deviation of power flow distribution at several frequencies (right)

is 0.25″, but decreases at 0.35″ and 0.45″. It is possible that wider ranges in measurement uncertainty will underestimate the participation of one or both of the closely spaced modes, resulting in lower variability in power flow at frequencies around these modes. Regardless, given that this frequency, and later shown at other frequencies near closely spaced modes, has the highest variability over the range of measurement uncertainties tested, it is clear that any error in measurements for the VPT can be greatly exacerbated at frequencies falling near closely spaced modes of the assembled system when calculating power flow.

The simulation results of Cases 2, 3, and 4 are shown in Fig. 1.5. The ranges for the perturbation parameters are given in Table 1.1. The normalized power flow and consistency distributions were visualized by assigning a color map to the distributions predicted from the Monte-Carlo simulation. The distributions were normalized by the maximum value of the empirical probability density distributions across the entire frequency range.

As previously mentioned, the distribution of power flow and consistency is not normally distributed across the entire frequency range. Additionally, the variability in power flow increases significantly at all frequencies around closely spaced modes of the assembled structures (approximately 1.2 kHz, 2.6 kHz, and 3.8 kHz), as was shown in Fig. 1.4 (right). The lower bound on the power flow distribution is truncated at −240 dB, which is only reached around 1 kHz. However, the power flow can still be negative at any of the frequencies shown. Ideally, the real part of the receiver virtual point mobility matrix is a positive semi-definite quantity because the net power flow into the receiver (and therefore power dissipated by the receiver) must be positive or zero. It is only zero in the case of no receiver damping, which is a numerical modelling insufficiency and does not occur in reality. The errors introduced by measurement uncertainty change the virtual point mobility such that the real part is no longer positive semi-definite, resulting in negative power flow.

The consistency is fairly high over the entire frequency range, only falling slightly below 0.9 at approximately 5.3 kHz. This corresponds to a slightly flexible interface near this frequency, but overall sensor consistency values greater than 0.9 are considered sufficient for the VPT rigid interface assumption to hold [7].

Looking at the effects of perturbing only the FRFs in Fig. 1.5 (middle), high variability in power flow is again observed near closely spaced modes. The distribution of power flow as a function of frequency looks similar to that of measurement uncertainty in the projection matrix across the frequency spectrum that was analyzed. However, the variability in power flow between 2.7 and 3.5 kHz is very small, suggesting that measurement uncertainties in the impact and response measurements do not significantly affect the calculation of power flow into the receiver between these frequencies.

Overall sensor consistency is also not greatly affected by measurement uncertainty in the FRFs. There is some slight variability above 5 kHz, but it is to be expected that there will be larger variability when the rigid interface assumption is invalid due to interface flexibility at a given frequency. For a more flexible interface, the overall sensor consistency would be lower, and a wider range of the consistency distribution would be expected. Unlike power flow, though, consistency is guaranteed to be bounded by 0 and 1 regardless of the measurement errors that propagate through its calculation because of its definition in Eq. 1.11.

Lastly, the effect of measurement uncertainties in both the projection matrix and FRFs was considered Fig. 1.5 (bottom). The distribution of power flow appears to be an average of the distributions obtained from Cases 2 and 3. However,

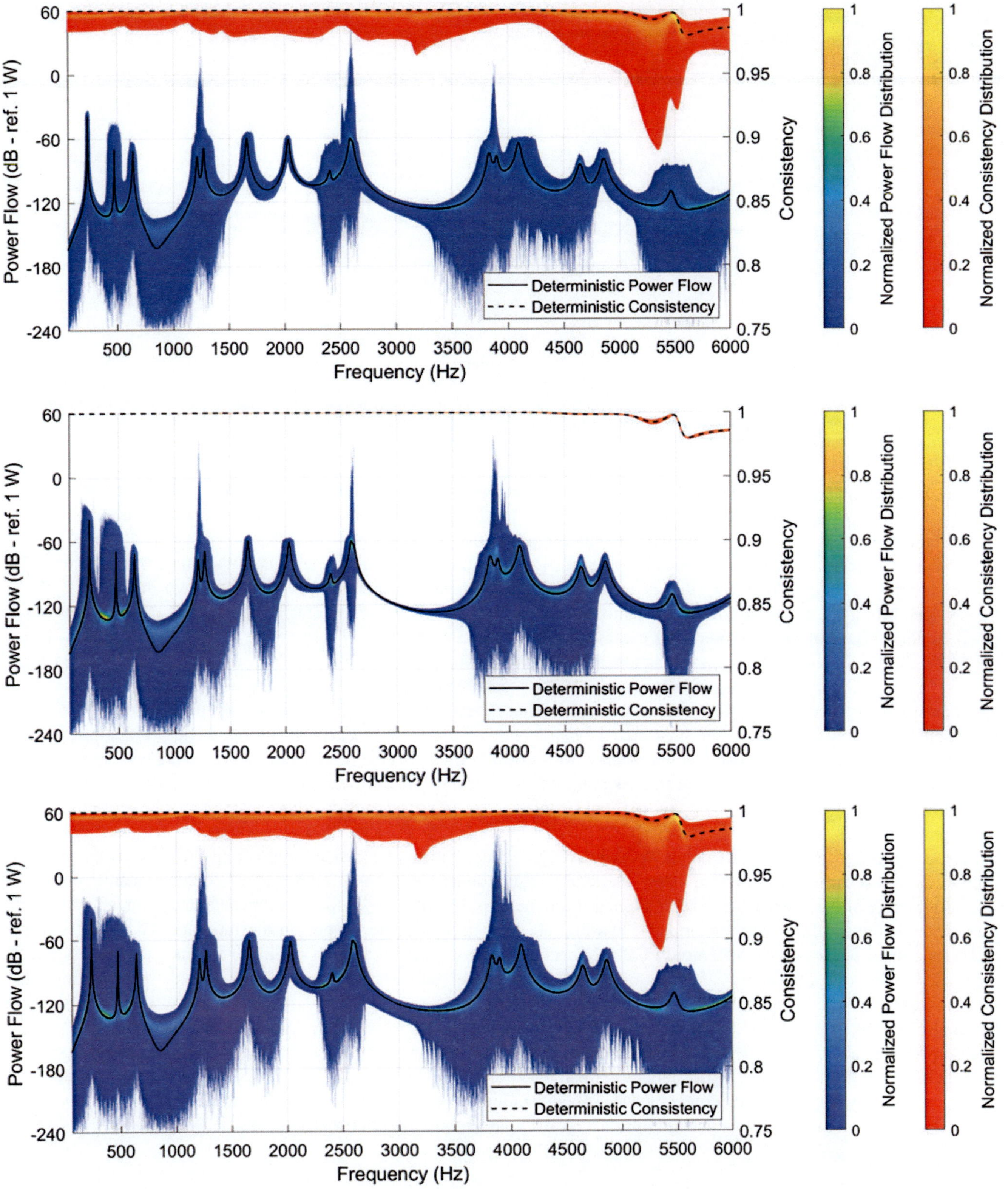

Fig. 1.5 Approximate normalized distribution of power flow (values less than −240 dB were truncated) and average VPT response consistency for uncertainty in the projection matrix (top), the interface FRFs (middle), and both the projection matrix and FRFs (bottom)

the distribution of consistency is heavily influenced by the measurement errors in the projection matrix. Since overall sensor consistency is a measure of the rigid interface assumption for the VPT at the locations of the accelerometers,

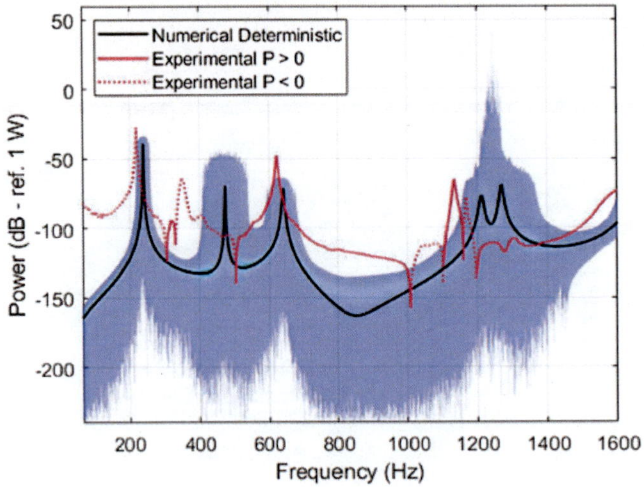

Fig. 1.6 Distribution of power flow for uncertainty in projection matrix and FRF measurements compared with power flow calculation using experimentally obtained data ($P > 0$ and $P < 0$ are positive and negative power flow in watts)

small measurement errors in the FRFs shouldn't greatly affect the consistency calculation. This is because these errors are present in both the numerator and denominator of Eq. 1.11 (both the physical and virtual point displacements), whereas the measurement errors in the projection matrix are only present in the numerator.

The numerical results obtained were compared to experimentally obtained data to determine if the range of uncertainties used was sufficient to bound the errors that accumulated in experimental measurements. This is shown in Fig. 1.6. The frequency spectrum analyzed was reduced to span from 60 to 1600 Hz. The experimentally calculated power flow is not bounded by the uncertainty ranges predicted using the Monte-Carlo simulation. There are several reasons for this. The first is uncertainty in impact and response locations was not taken into account due to the fine interface mesh required for that study. Only uncertainty in the projection matrix distance measurements, containing information regarding the location of the accelerometers and applied forces relative to the virtual point, was considered. Physically moving the force and accelerometer locations will change the measured FRFs, leading to errors in the actual FRFs and the projection matrix measurements. Second, a considerable amount of mass was added to the structure when the accelerometers were secured to the interface. These were modelled as point masses in the Monte-Carlo simulation, but their geometric size was comparable to the geometric size of the interface, and the rotational effects of the interface motion could not properly be accounted for in the numerical model.

Finally, a significant amount of damping, due to the high concentration of localized mass from the accelerometers on the interface, was added to the structure when interface measurements were made. This was observed acoustically using the ring-down time after an impact had been applied to the beams. The ring-down time was substantially longer when no accelerometers were on the interface compared to when all three accelerometers were on the interface. Given that most previous successful studies on the VPT used PCB-B21 accelerometers, which are much smaller in geometry and mass than the ones used in this study, the effects of accelerometer mass loading on the interface play a significant role in predicting the dynamics of the assembled structures.

It can also be observed in Fig. 1.6 that the power flow is negative at some locations in the frequency spectrum analyzed. The reason for this could be due to any of the aforementioned shortcomings of the experiment. The deterministic power flow – that is, power flow calculated with no measurement uncertainty – in the projection matrices or FRFs is not an accurate predictor of the experimental power flow. However, understanding how certain types of errors in the VPT can propagate through the calculation of power flow is still important to an engineer who might be applying the VPT in practice.

1.6 Conclusion

A Monte-Carlo simulation was performed by perturbing the elements of the VPT projection matrix and interface FRFs. It was shown that the distribution of power flow is not Gaussian, and it is most variable near frequencies with closely spaced modes. The variability in power flow as a function of frequency was similar for uncertainty in the projection matrix quantities and for the FRFs. Additionally, the overall sensor consistency is more dependent on measurement errors in the projection matrix when compared to measurement errors in the angles of the impacts and accelerometers. The numerical results were compared to experimentally obtained data which was used to calculate power flow, and the range of uncertainties tested did not appear to account for the experimental errors present. The dynamics of the interface with the added mass of the accelerometers could also not be properly modelled due to the large size of the accelerometers relative to the interface. From these results, it should be emphasized that if the VPT is to be used to assemble two substructures and calculate power flow between them, the accuracy in the measurement of the distance and orientation parameters of the projection matrices and of the FRFs is of primal importance, as errors can lead to large variability in the calculation of power flow.

Acknowledgments The authors would like to thank the Walker Graduate Assistantship for funding this research.

References

1. de Klerk, D., Rixen, D.J., Voormeeren, S.N.: General framework for dynamic substructuring: history, review and classification of techniques. AIAA J. **46**(5), 1169–1181 (2008)
2. Jetmundsen, B., Bielawa, R.L., Flannelly, W.G.: Generalized frequency domain substructure synthesis. J. Am. Helicop. Soc. **33**(1), 55–64 (1988)
3. de Klerk, D., Rixen, D.J., de Jong, J.: The frequency based substructuring (FBS) method reformulated according to the dual domain decomposition method. In: Proceedings of the 24th International Modal Analysis Conference, A Conference on Structural Dynamics (2006)
4. Van der Valk, P.L.C., Van Wuijckhuijse, J.B., De Klerk, D.: A benchmark test structure for experimental dynamic substructuring. In: Structural Dynamics, vol. 3, pp. 1113–1122. Springer, New York (2011)
5. Trainotti, F., Berninger, T.F.C., Rixen, D.J.: Using laser vibrometry for precise FRF measurements in experimental substructuring. In: Dynamic Substructures, vol. 4, pp. 1–11. Springer, Cham (2020)
6. Pasma, E.A., et al.: Frequency based substructuring with the virtual point transformation, flexible interface modes and a transmission simulator. In: Dynamics of Coupled Structures, vol. 4, pp. 205–213. Springer, Cham (2018)
7. van der Seijs, M. Experimental Dynamic Substructuring: Analysis and Design Strategies for Vehicle Development (2016)
8. Allen, M.S., et al.: Substructuring in Engineering Dynamics. Springer, Cham (2020)
9. van der Seijs, M.V., et al. An improved methodology for the virtual point transformation of measured frequency response functions in dynamic substructuring. In: 4th ECCOMAS Thematic Conference on Computational Methods in Structural Dynamics and Earthquake Engineering, No. 4 (2013)
10. Barten, E., van der Seijs, M.V., de Klerk, D.: A complex power approach to characterise joints in experimental dynamic substructuring. In: Dynamics of Coupled Structures, vol. 1, pp. 281–296. Springer, Cham (2014)

Chapter 2
Using Flight Test Measurements on a Low-Fidelity Component to Predict Response on a High-Fidelity Component

Randall L. Mayes

Abstract Measurements are often made during system field tests on a component of interest and the system. But the excitation forces are unknown. These tests are early in the development phase so the component may be of low fidelity. The environmental response of the high-fidelity component is what is desired to develop appropriate component qualification specifications. In this paper, the pseudo-force method from the field of transfer path analysis (TPA) is utilized in an analytical rocket and component example. A tractable number of frequency response function (FRF) measurements are made on the rocket and low-fidelity component. The low-fidelity component on a fixture is characterized by laboratory FRFs. Then the rocket and component are subjected to a random vibration loading in a flight test, with accelerations measured on the component and its base. Presumably at a later time, when the high-fidelity component is available, it is characterized with FRFs at the field measurement locations. The pseudo-force method is utilized to predict the flight response of the high-fidelity component without an additional flight test. A simple transmission simulator (fixture) is utilized in this work to minimize experimental errors at connection degrees of freedom (DOF).

Keywords Flight test · Transfer path analysis · Smart dynamic testing · Pseudo-force

2.1 Motivation

System flight tests are often conducted with vibration measurements on low-fidelity components before the high-fidelity component has been completely developed. The measurements are utilized for developing qualification testing specifications for the components. However, there is an often unquantified leap of uncertainty from the low-fidelity component response and the high-fidelity component specification. Since flight tests are quite expensive, there is often no flight test performed with the high-fidelity component. Ideally, specification engineers would like to have the measurements from the flight with the high-fidelity component. Transfer path analysis (TPA) provides a method to predict the response of the high-fidelity component with certain measurements and analysis. In this work, an analytical rocket system with a low-fidelity component is utilized to demonstrate that the pseudo-force method of TPA can theoretically be used with frequency-based substructuring to predict the response the high-fidelity component would have had in flight (without the additional flight test).

2.2 Analytical Problem

A planar analytical model of a 20-foot-long rocket is formulated with a base mounted component set off center in the forward section. With this model, it is assumed that some FRFs can be measured on the component and near its attachment due to force inputs on the nose, near the aft end, and on the component plate. The flight is simulated with a random axial force applied to the nozzle of 1000 lbf RMS and with a random lateral force of 100 lbf RMS. Simulated measurements are made during the flight on the low-fidelity component. Simulated measurements of FRFs from some rocket input locations to the component measurement locations are utilized with the flight measurements on the component to calculate pseudo-forces that exactly simulate the response on the component. Both the low-fidelity and high-fidelity components are characterized with FRFs from a base mounted fixture to the component response DOF. Frequency-based substructuring is utilized to switch

R. L. Mayes (✉)
Consultant, Albuquerque, NM, USA

M. Allen et al. (eds.), *Dynamic Substructures, Volume 4*, Conference Proceedings of the Society for Experimental Mechanics Series,
https://doi.org/10.1007/978-3-031-04094-8_2

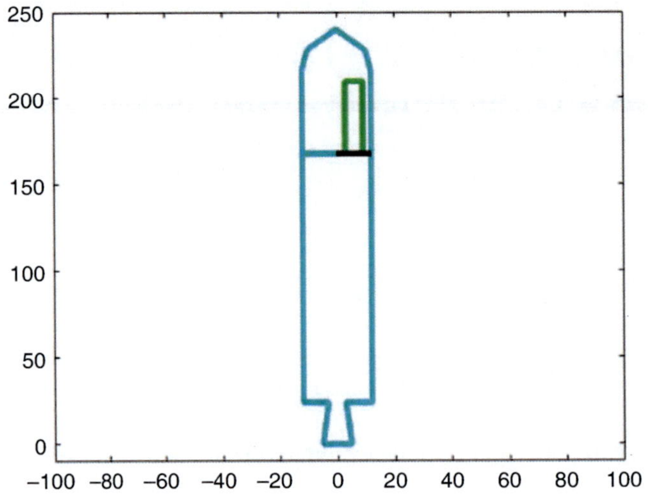

Fig. 2.1 Representation of the rocket and component

out the low-fidelity component with the high-fidelity component on the rocket. Then the pseudo-forces are applied to the revised rocket-component model FRFs to predict the high-fidelity component response in the flight. Figure 2.1 represents the rocket/component model.

2.3 Pseudo-force Theory and Result

The pseudo-force method of TPA is presented by Van der Seijs in a text [1]. The flight response is caused by the forces on the nozzle. If we apply a non-unique set of pseudo-forces that perfectly cancels the motion at the rocket/component interface (black line in Fig. 2.1), then if the operational forces are removed, the pseudo-forces would cause the same response as the flight at the interface. If the interface motion is correct, then the motion on the component will also be correct. In this case, we develop FRFs between the lateral and vertical forces at the two blue locations and the red and purple locations on the component as shown in Fig. 2.2. Knowing the flight responses at the red locations on the component, the pseudo-forces are calculated in the frequency domain from Eq. (2.1) as

$$\left\{pf_{\text{blue}}\right\} = \left[Y_{\text{red,blue}}\right]^{+}\left\{u_{\text{red}}\right\} \tag{2.1}$$

where the vector of pseudo-forces is denoted as pf_{blue}, the FRFs between the red component responses and the pseudo-forces are $Y_{\text{red,blue}}$, and the measured flight low-fidelity component responses are u_{red} as shown in Fig. 2.2. The pseudo-force calculation must be over-determined.

For this case, there are four pseudo-force (lateral x and vertical y) inputs at the circled blue locations calculated from the eight lateral x and vertical y responses at all the component red circled locations. After performing the indicated calculations, the FFT response for the x direction at one of the red circled locations is given in Fig. 2.3 for both the original flight forcing function and the pseudo-forces. In Fig. 2.4 is the same type of plot for the lateral response at the purple circled location which was not used in the pseudo-force calculation of Eq. (2.1).

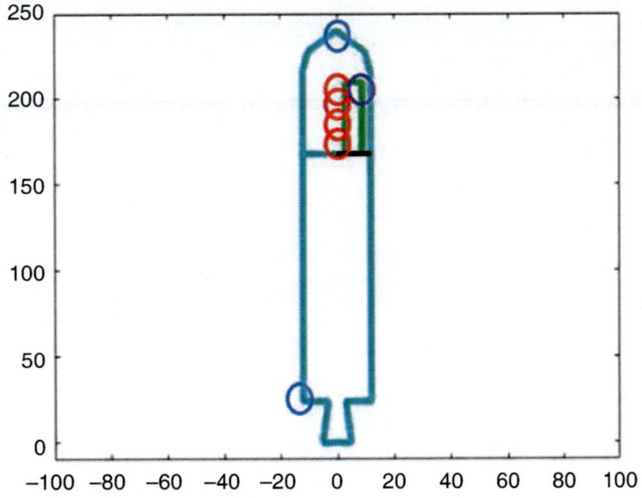

Fig. 2.2 Rocket pseudo-force locations in blue, component response locations for determining pseudo-force in red, and truth location in purple

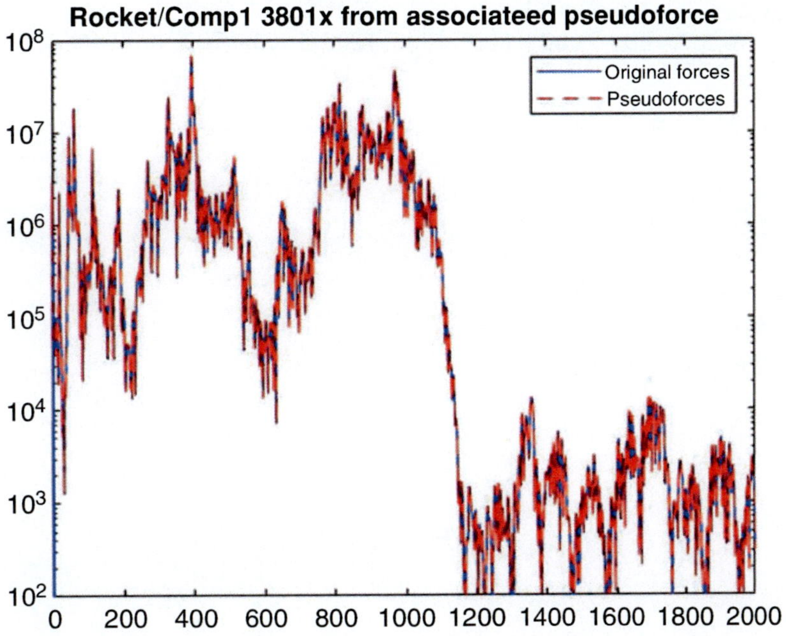

Fig. 2.3 FFT response at 3801X (blue is from the nozzle forces and red is from the four pseudo-forces)

2.4 Component Measurements

In the substructuring approach described later, there are three sets of FRFs required to complete the analysis. One set will be on the low-fidelity component mounted in the rocket, one set will be on the low-fidelity component mounted on a fixture, and the third set will be on the high-fidelity component mounted on the same fixture. All three sets require these measurements. In Fig. 2.5, the measurement locations are shown, with both lateral and vertical DOF required at each location. The responses at all red, blue, and purple locations are required due to force inputs at all the circled blue locations. In addition, the rocket set will require the responses at all 18 DOF of Fig. 2.5 from the 4 pseudo-force input locations. The low-fidelity component FRFs will be subtracted from the rocket/component FRFs, and the high-fidelity FRFs will be added with appropriate constraints to develop the desired rocket/high-fidelity component FRFs.

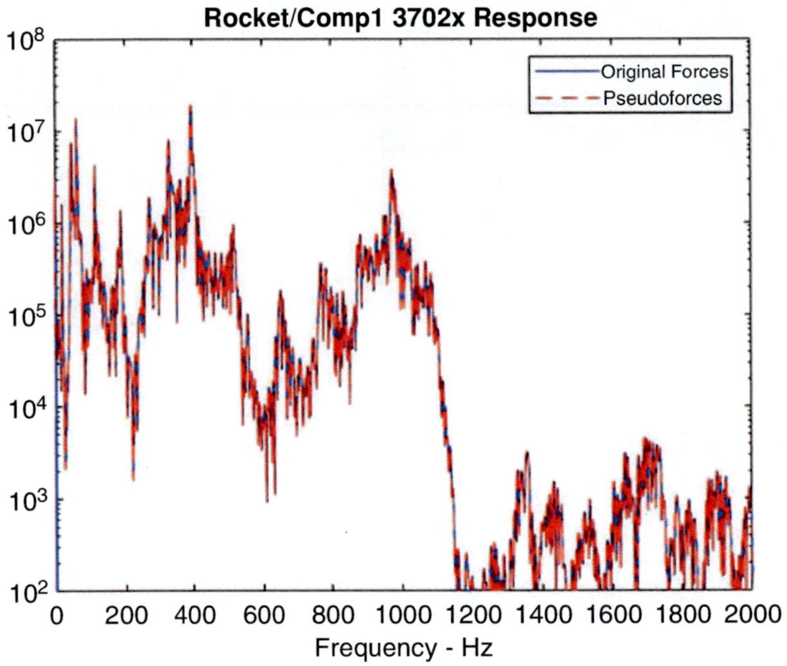

Fig. 2.4 FFT response at 3702X (blue is from the nozzle forces and red is from the four pseudo-forces). This response was not utilized in pseudo-force calculations

Fig. 2.5 Closeup of component measurement locations

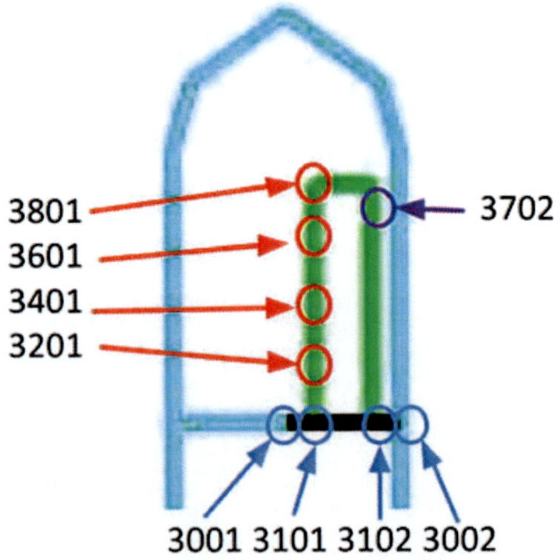

2.5 Frequency-Based Substructuring with Transmission Simulator

The frequency-based substructuring (FBS) theory using the dual formulation is provided by Rixen in a textbook [2]. Portions are reproduced here. The frequency domain equations of motion can be written as

$$\left[K + j\omega C - \omega^2 M\right]\{u\} = \{f\} + \{g\} \tag{2.2}$$

where u are the displacement DOF, f are the external forces, and g are the connection forces between substructures. The stiffness, damping, and mass matrices are block diagonal corresponding to the matrices of each substructure, e.g.,

$$K = \begin{bmatrix} K_1 & 0 & 0 \\ 0 & -K_2 & 0 \\ 0 & 0 & K_3 \end{bmatrix} \tag{2.3}$$

where the rocket/low-fidelity component would be the first substructure, the low-fidelity component on fixture would be the second substructure that we are going to subtract, and substructure 3 would be the high-fidelity component on fixture that we are going to add. The constraints between connected DOF in different substructures can be written with a Boolean matrix of 1,0,−1 values as

$$[B]\{u\} = \{0\} \tag{2.4}$$

and the connection forces can be written so that they are in equilibrium as

$$\{g\} = -[B]^T\{\lambda\} \tag{2.5}$$

where a specific λ is the absolute value of the force between two connection DOF. Inverting the uncoupled dynamic matrix to calculate the uncoupled FRF matrix Y and solving for λ give the FBS equation as

$$\{u\} = \left[Y - YB^T\left[BYB^T\right]^{-1}BY\right]f \tag{2.6}$$

in which the Y matrix is the uncoupled block diagonal matrix of the different substructure FRF matrices, analogous to the K matrix of Eq. (2.3). If there are more input DOF than output DOF, the physical B matrix will be different because the list of u DOF is different for the inputs than the outputs. For this case in Eq. (2.6), the output B matrix is used wherever there is a B, and the input B matrix is used wherever the B matrix is transposed.

2.6 FRF Component Subtraction and Addition with Transmission Simulator

Substructures 2 and 3, the low-fidelity and high-fidelity components, are attached to a fixture. We utilize this fixture as the transmission simulator (TS) as introduced by Allen and Mayes [2]. In this case, the simplest TS is utilized with the assumption that the TS is rigid. In our planar problem, that means that there will be three rigid body mode shapes of the TS, vertical, lateral, and pitch. The TS mode shape matrix will be formed for all eight connection DOF at nodes 3001, 3002, 3101, and 3102 of Fig. 2.5 and denoted as Ψ. The physical constraint matrix, B, has eight constraints coupling substructures 1 and 2 and eight constraints coupling substructures 2 and 3. Soften the constraint matrix from 16 down to 6 constraints by premultiplying each set of eight constraints in Eq. (2.4) by the pseudoinverse of Ψ, yielding a modified matrix B_{TS}, which is utilized to replace B in Eqs. (2.4) and (2.6).

One may ask what is the advantage of the relaxed TS constraint over the regular physical constraint. It is known that the experimental errors from measured FRFs will cause ill-conditioning and spurious modes in FBS. The relaxing of the constraint provides a least squares compatibility between substructures mitigating the ill-conditioning. As long as the TS mode shapes approximately span the motion at the attachment DOF, little accuracy is lost in the coupling calculation. For this analysis, eight attachment DOF were used on all three substructures. However, if one utilizes the TS constraint, any number of additional forcing and response DOF on the fixture (or rocket component plate) could be added to help characterize the attachments between substructures, which may be of value in certain situations when all the drive point measurements may be difficult to obtain on the connection hardware.

Note that the second block of Eq. (2.6) will be $-Y$ of the low-fidelity component substructure since we are subtracting it. The addition of the high-fidelity component and subtraction of the low-fidelity component from the rocket/low-fidelity component is accomplished by calculating the new FRFs within the brackets of Eq. (2.6) with B_{TS} replacing the physical B.

2.7 Final Prediction of Flight with High-Fidelity Component

After the coupled FRF matrix is calculated, the pseudo-forces are applied in Eq. (2.6) and the responses calculated at the appropriate DOF on the high-fidelity component. In Figs. 2.6 and 2.7, one can see the absolute value of the FFTs of the response for DOF 3702 x and y (see location in Fig. 2.5). The blue response is based on the original nozzle forces being applied to the rocket with the high-fidelity component (the truth response), and the red response is the one calculated using the pseudo-forces and the FRFs after the subtraction of the low-fidelity component and addition of the high-fidelity component.

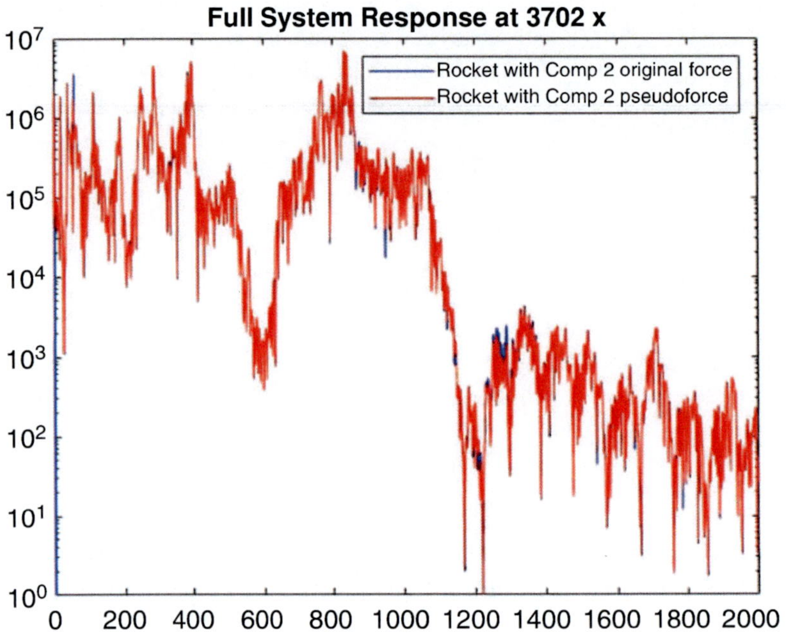

Fig. 2.6 Response at 3702x on high-fidelity component in rocket

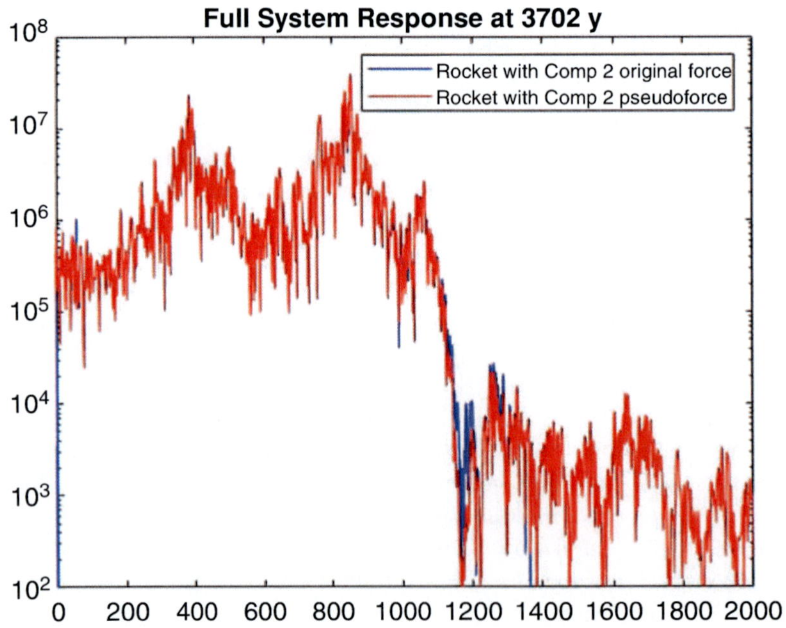

Fig. 2.7 Response at 3702Y on high-fidelity component in rocket

For a view of the differences between the flight response between the low-fidelity component and the high-fidelity component in the rocket, see Figs. 2.8 and 2.9. At some frequencies, there is close to an order of magnitude difference between the low-fidelity and high-fidelity component responses.

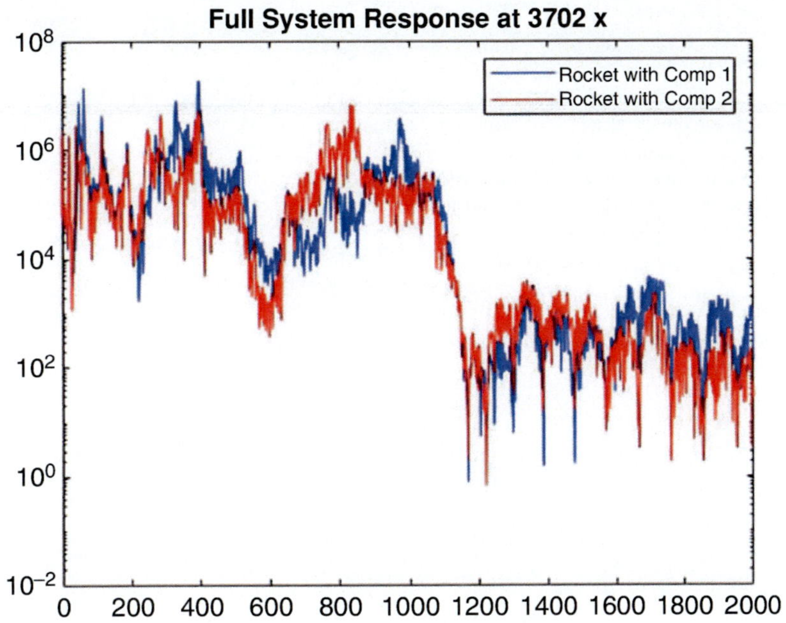

Fig. 2.8 Response of low-fidelity component in rocket (blue) vs. high-fidelity component (red) at DOF 3702x

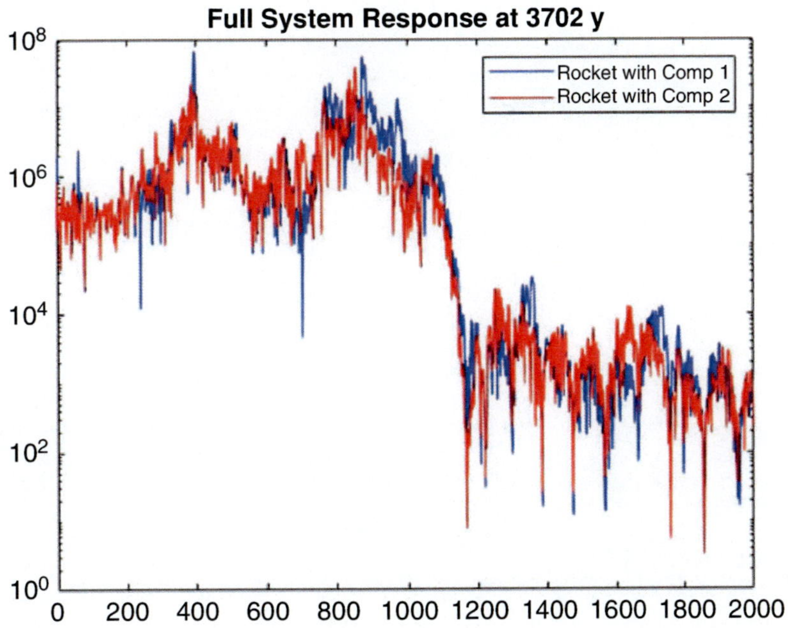

Fig. 2.9 Response at DOF 3702y of low-fidelity component in rocket (blue) vs. high-fidelity component (red)

2.8 Conclusions

Using analytical models, this work demonstrates a theoretical approach to correcting responses on a low-fidelity component on a rocket flight test to estimate what the responses would have been on a high-fidelity component developed after the flight test. This could provide environments engineers with higher-confidence measurements to improve the process of deriving vibration qualification specifications for high-fidelity components and their components without additional expensive flight testing. The method utilizes the pseudo-force approach from transfer path analysis along with frequency-based substructuring to replace the low-fidelity component with the high-fidelity component. The transmission simulator constraint is proposed to mitigate coupling errors due to typical experimental errors inherent in FRF measurements.

References

1. Allen M.S., Rixen D., van der Seijs M., Tiso P., Abrahamsson T., Mayes R.L.: Chapter 5.4.6, Component TPA: pseudo-forces. In: Substructuring in Engineering Dynamics – Emerging Numerical and Experimental Techniques, Volume 594 of CISM International Centre for Mechanical Sciences, pp. 199–201. Springer, Cham (2020). https://doi.org/10.10079/978-3-030-25532-9
2. Allen M.S., et al.: Chapter 2, Preliminaries: primal and dual assembly of dynamic models, Chapter 4, Experimental substructuring. In: Substructuring in Engineering Dynamics – Emerging Numerical and Experimental Techniques, Volume 594 of CISM International Centre for Mechanical Sciences. Springer, Cham (2020). https://doi.org/10.10079/978-3-030-25532-9

Chapter 3
Road Noise NVH Part 2: Exploring the Capabilities of the TPA Framework with Interface Forces

Julie M. Harvie, Maarten V. van der Seijs, David P. Song, and Munhwan Cho

Abstract In the previous paper (van der Seijs MV, Harvie JM, Song DP, Road noise NVH: embedding suspension test benches in NVH design using Virtual Points and the TPA framework. In: Proceedings of the thirty-ninth international modal analysis conference, 2021), we presented the feasibility of using a tire noise test bench in the engineering of full-vehicle NVH. Road noise at the driver's ear was accurately simulated with blocked forces obtained using several different methods: with the suspension on two test benches (one rigid and one compliant) and in the full vehicle. All methods produced accurate results, using virtual points in DIRAC to ensure compatibility between the various test assemblies.

In this paper, we further investigate the capabilities of the framework using interface forces instead of blocked forces. Several methods are used to calculate interface forces: a matrix inverse method, a mount stiffness method, and with interface forces derived from blocked forces. Results are evaluated for the simulation of road noise at the driver's ear in the vehicle.

Keywords NVH · Blocked forces · Transfer path analysis · Virtual point

Nomenclature

$\mathbf{u}$	Dynamic displacements/rotations
$\mathbf{f}$	Applied forces/moments
$\mathbf{g}$	Interface forces/moments
$\mathbf{Y}$	Admittance FRF matrix
$\mathbf{Z}^{\mathrm{mt}}$	Mount stiffness matrix
$\star^{\mathrm{AB}}$	Pertaining to the assembled system
$\star^{\mathrm{A}}$	Pertaining to the active component
$\star^{\mathrm{B}}$	Pertaining to the passive component
$\star^{\mathrm{R}}$	Pertaining to the test rig
$\star_1$	Source excitation DoF
$\star_2$	Interface DoF
$\star_3$	Receiver/target DoF
$\star_4$	Indicator DoF (around the interface, for FRF over-determination)

J. M. Harvie (✉)
Self, Leominster, MA, USA

M. V. van der Seijs
VIBES.technology, Delft, The Netherlands
e-mail: mvanderseijs@vibestechnology.com

D. P. Song · M. Cho
Hyundai Motor Company – NVH Research Lab, Hwaseong-si, Gyeonggi-do, South Korea
e-mail: dsong@hyundai.com; munhcho@hyundai.com

M. Allen et al. (eds.), *Dynamic Substructures, Volume 4*, Conference Proceedings of the Society for Experimental Mechanics Series,
https://doi.org/10.1007/978-3-031-04094-8_3

3.1 Introduction

Road noise is known to be one of the major noise sources in a modern car. While drivetrain noise levels nowadays go down considerably – already due to replacement of the combustion engine by an electric one – noise induced by the tire-road contact remains an important NVH aspect. For a typical consumer car, the structure-borne noise contribution sits predominantly in the frequency range from 20 to 500 Hz, whereafter the airborne contributions gradually take over.

A major point of focus for an OEM to mitigate road noise is the suspension subsystem, comprised of the knuckle-wheelhub assembly, spring/damper, and various linkages. Figure 3.1 shows the topology of a typical rear suspension (the one studied in this project), highlighting the interfaces between the suspension and vehicle body; these interfaces are further discussed in the "Approach" section.

Although not fully emphasized in this diagram, the operational loads for road noise come into the system through the knuckle-wheelhub assembly, travel through the suspension system and vehicle body, and ultimately generate noise at the driver's ear. Thus, the engineer's design space for NVH optimization primarily includes the dynamic properties of the linkages and spring/damper and, perhaps more importantly, the stiffness of the rubber bushings that connect many of these linkages. However, each bushing and component must be carefully designed not only to optimize the vehicle's NVH performance but to balance its performance for handling, drive comfort, and NVH performance. Thus, the optimization of the suspension for NVH performance is clearly a multi-variable challenge and therefore needs a process that can be used to evaluate changes on the full-vehicle level.

3.2 Motivation

In the previous paper [1], a thorough study of the efficacy of component-based transfer path analysis (TPA) methods [2] was performed. That study involved using blocked forces to estimate the loads imparted by the rear suspension system onto the vehicle body during a typical run-up. Using blocked forces as the quantity of interest provides us with the possibility of applying the loads to any other vehicle with that suspension (including not only other prototypes but conceptual designs as well).

While very accurate results were observed with the component-based TPA approach, other TPA techniques could also be used to analyze the system. While the blocked forces clearly provide an excellent basis for capturing loads early in the

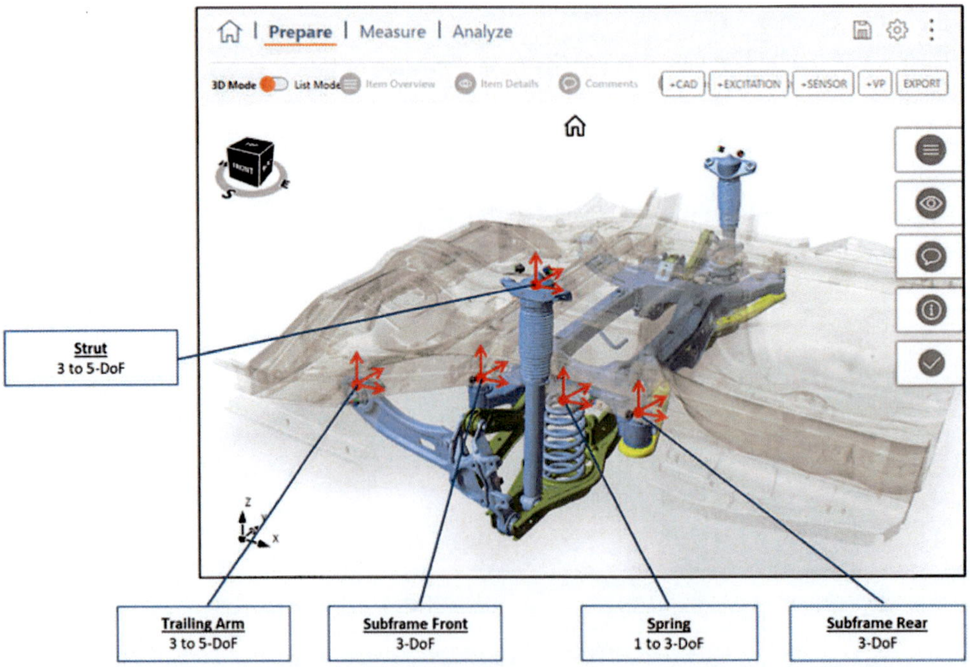

Fig. 3.1 The five connection points of the rear suspension and their modelling strategy using virtual points

design cycle, it may be desired later in the design cycle to understand the "true" interface forces between components. Using interface forces as the quantity of interest, as is done in classical TPA, also allows for follow-on activities such as power flow analysis [3]. Another potential advantage of using interface forces is that they are combined with trimmed body dynamics rather than requiring full-vehicle dynamics for TPA predictions. Thus, in this paper, we present results from classical TPA analyses, complementary to the component-based TPA results shown in the previous paper.

3.3 Approach

For a thorough review of both classical and component-based TPA methods, the reader is referred to [2]. An overview is presented here using the same nomenclature, as highlighted in Fig. 3.2. Note that in the following equations, the time/frequency dependency of the variables is dropped for simplicity.

First, the modelling strategy for the connection points is discussed, as this is used throughout all studies. Then, a brief description of the three interface force methods is given, followed by information about the trimmed body tests.

3.3.1 Connection Points

The compatibility of vehicle transfer functions, trimmed body transfer functions, and test rig forces is critical to the successful application of the methods described below. In this study, as well as in the previous paper [1], we use virtual points [4] at prescribed locations to have a common interface between all test assemblies. Data is obtained at the virtual points with ease in DIRAC, which allows the engineer to make traditional FRF/NTF measurements and transform either the forces, responses, or both to the virtual points.

Figure 3.1 showed the five connection points of the rear suspension to the vehicle bodywork: the trailing arm, subframe front and rear mounts, strut with damper, and the spring. In total, we consider the ten coupling points for the left and right side simultaneously.

As indicated, the modelling strategy can be chosen for each coupling point differently. The four subframe mounts typically only exhibit translational motion and can be well described by 3-DoF translational dynamics. For the trailing arm and the strut, additional rotational DoF have been added for optional moment information.

3.3.2 Interface Forces: Matrix Inverse Method

In the first of the three methods for capturing interface forces, a matrix inverse is employed with:

$$\mathbf{g}_2 = \left(\mathbf{Y}_{42}^{\mathbf{B}}\right)^+ \mathbf{u}_4 \tag{3.1}$$

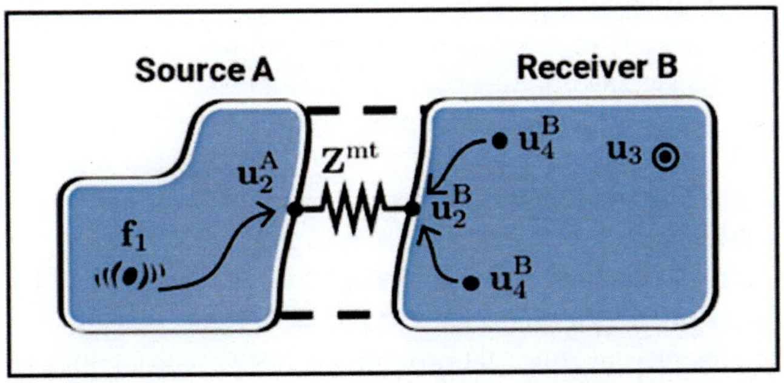

Fig. 3.2 Overview of TPA framework terminology

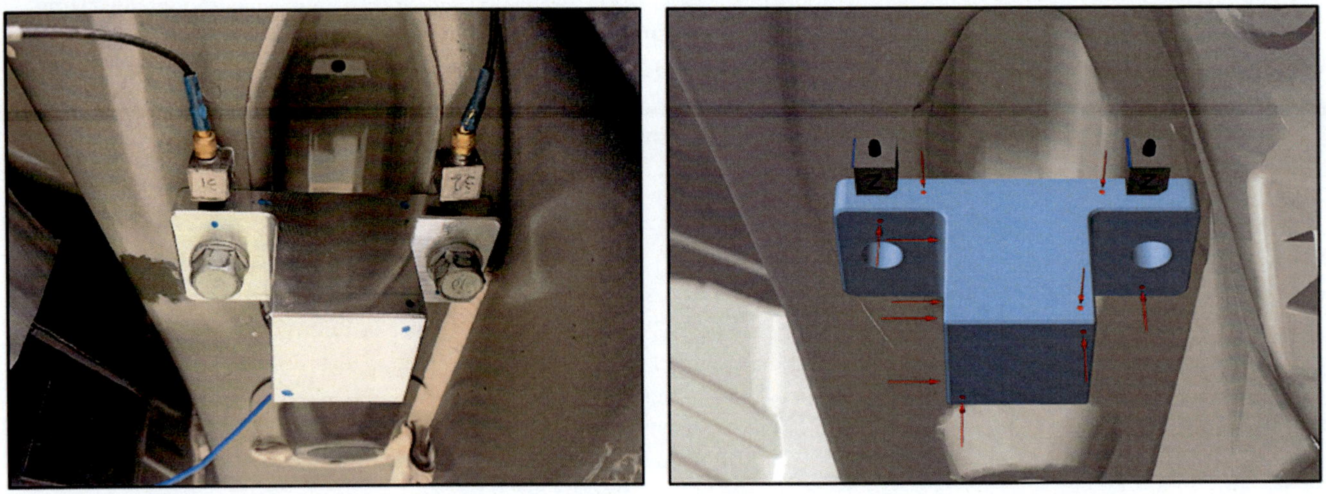

Fig. 3.3 Example setup at the trailing arm for FRF measurements on the trimmed body for matrix inverse calculation, (*left*) physical test setup and (*right*) virtual setup in DIRAC

Essentially, this involves recording displacements (or more commonly accelerations) near the interfaces during the vehicle operation ($\mathbf{u}_4$), removing the suspension, and measuring the FRFs ($\mathbf{Y}^{\mathbf{B}}_{42}$) between those accelerometers and impacts at the interfaces. Inverting the trimmed body FRFs and multiplying with the accelerations give us the interface forces, $\mathbf{g}_2$.

Figure 3.3 shows an example of the setup used to obtain $\mathbf{Y}^{\mathbf{B}}_{42}$ on the trimmed body in this project, including the accelerometers and impacts at the trailing arm interface. With the suspension removed, transmission simulators [5] are installed in its place to mimic the suspension interfaces and allow the engineer to excite all 6-DoF with a typical impact hammer. Here, we see how the virtual point transformation must be applied to the impacts in order to get forces and moments exactly at the interfaces. In this example, three indicator accelerometers are installed on the body side, and the virtual point transformation is not used; this over-determination of the FRF matrix is essential for an accurate matrix inverse.

In the corresponding full-vehicle operation test, not shown here, the accelerometers were installed in identical locations and orientations to those shown in Fig. 3.3.

3.3.3 Interface Forces: Mount Stiffness Method

Another technique for obtaining interface forces is the mount stiffness method. This method has the advantage that a matrix inverse is not required, thus eliminating the need to consider matrix conditioning, etc. Using this method, the interface forces $\mathbf{g}_2$ are calculated using:

$$\mathbf{g}_2 = \mathbf{Z}^{\mathrm{mt}} \left(\mathbf{u}^{\mathbf{A}}_2 - \mathbf{u}^{\mathbf{B}}_2 \right) \tag{3.2}$$

where the motion on the active and passive side of flexible mounts are captured in $\mathbf{u}^{\mathbf{A}}_2$ and $\mathbf{u}^{\mathbf{B}}_2$, respectively, and the relative motion is combined with the (frequency-dependent) stiffness of all bushings, $\mathbf{Z}^{\mathrm{mt}}$. This method and setup is illustrated in Fig. 3.4 and works best for determining interface forces at connection points with resilient mounts, where the stiffness of those mounts is known with high accuracy.

In this project, the inverse substructuring technique for mount characterization [6] was used to determine the dynamic stiffness of all mounts; an example is shown for a subframe mount in Fig. 3.5. As seen, this technique involves attaching fixtures to both the active and passive sides of the bushings and measuring the full 12x12 FRF matrices of the system (three translations and three rotations per fixture). This is achieved using accelerometers and impact hammers with the virtual point transformation in DIRAC. We simply double-integrate and invert the measured accelerance matrix, and assuming a specific "DoF to DoF" topology, the transfer stiffness is contained in the off-diagonal terms:

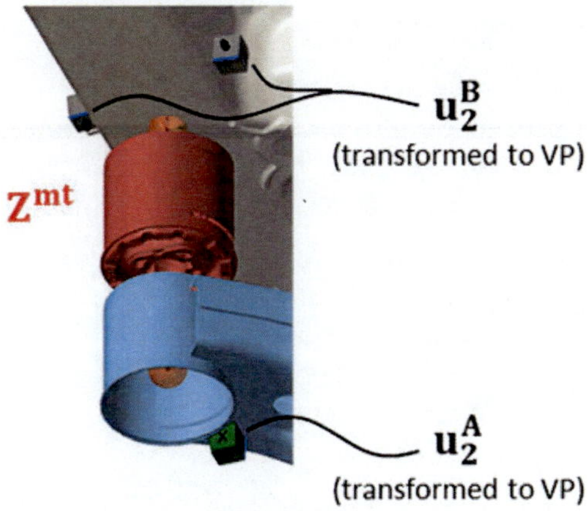

Fig. 3.4 Illustration of mount stiffness method

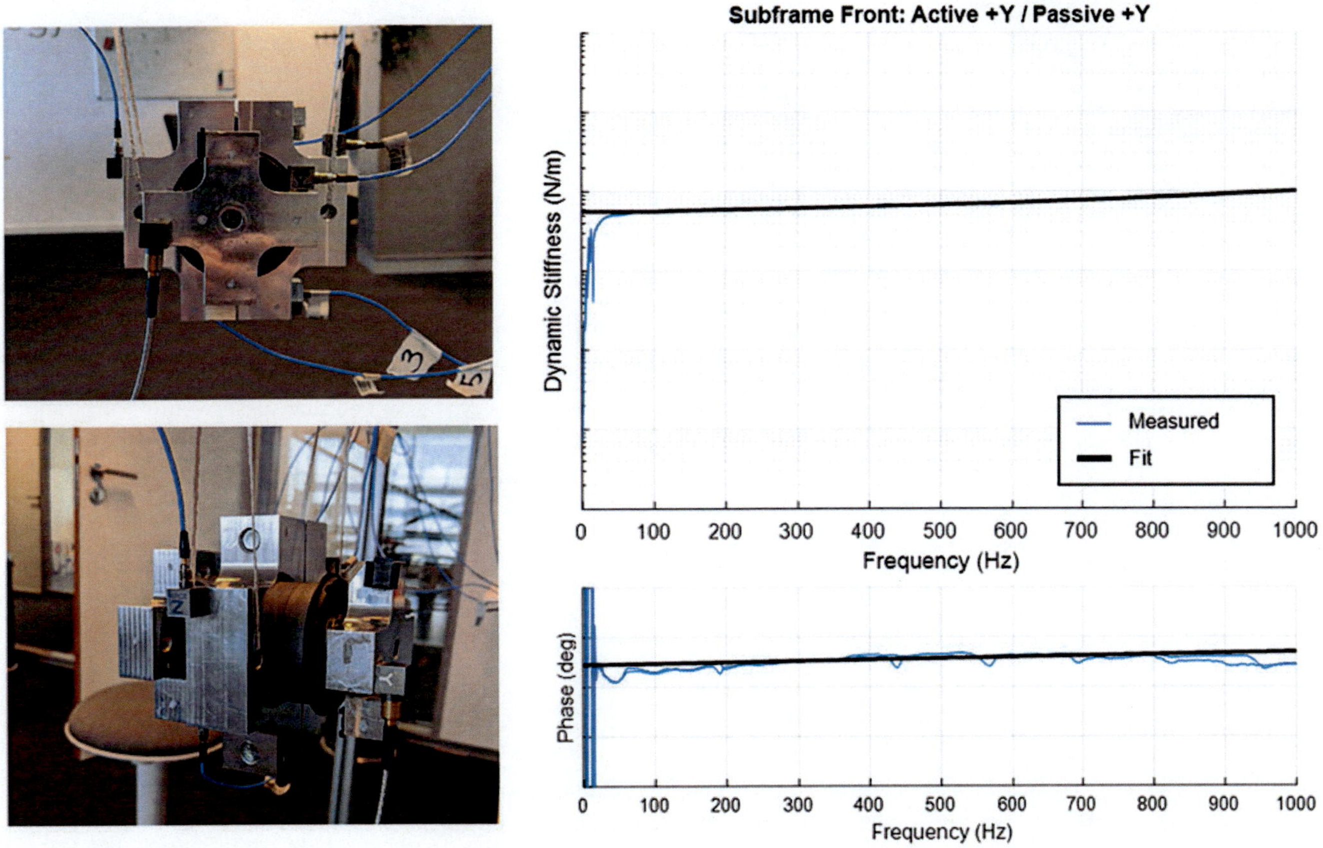

Fig. 3.5 Example of dynamic stiffness characterization for the subframe mount; (*left*) test setup and (*right*) stiffness results (both measured and fit)

$$\left(\mathbf{Y}^{\mathrm{AIB}}\right)^{-1} = \mathbf{Z}^{\mathrm{AIB}} = \begin{bmatrix} \mathbf{Z}_{11}^{\mathrm{A}} + \mathbf{Z}_{11}^{\mathrm{I}} & \mathbf{Z}_{12}^{\mathrm{I}} \\ \mathbf{Z}_{21}^{\mathrm{I}} & \mathbf{Z}_{22}^{\mathrm{B}} + \mathbf{Z}_{22}^{\mathrm{I}} \end{bmatrix} \tag{3.3}$$

Isolator stiffness

As seen in Fig. 3.5, for an adequate test (easily judged using the quality indicators in DIRAC), the resulting transfer stiffness often follows a flat line in the mid-frequency band. Although not shown, the extracted stiffness values of the bushings matched remarkably well with the vendor-supplied static stiffnesses.

It is worth mentioning here that only six out of ten interfaces defined in Fig. 3.1 include resilient mounts, and therefore interface forces were only calculated for those six connection points (the four subframe mounts and the two trailing arms).

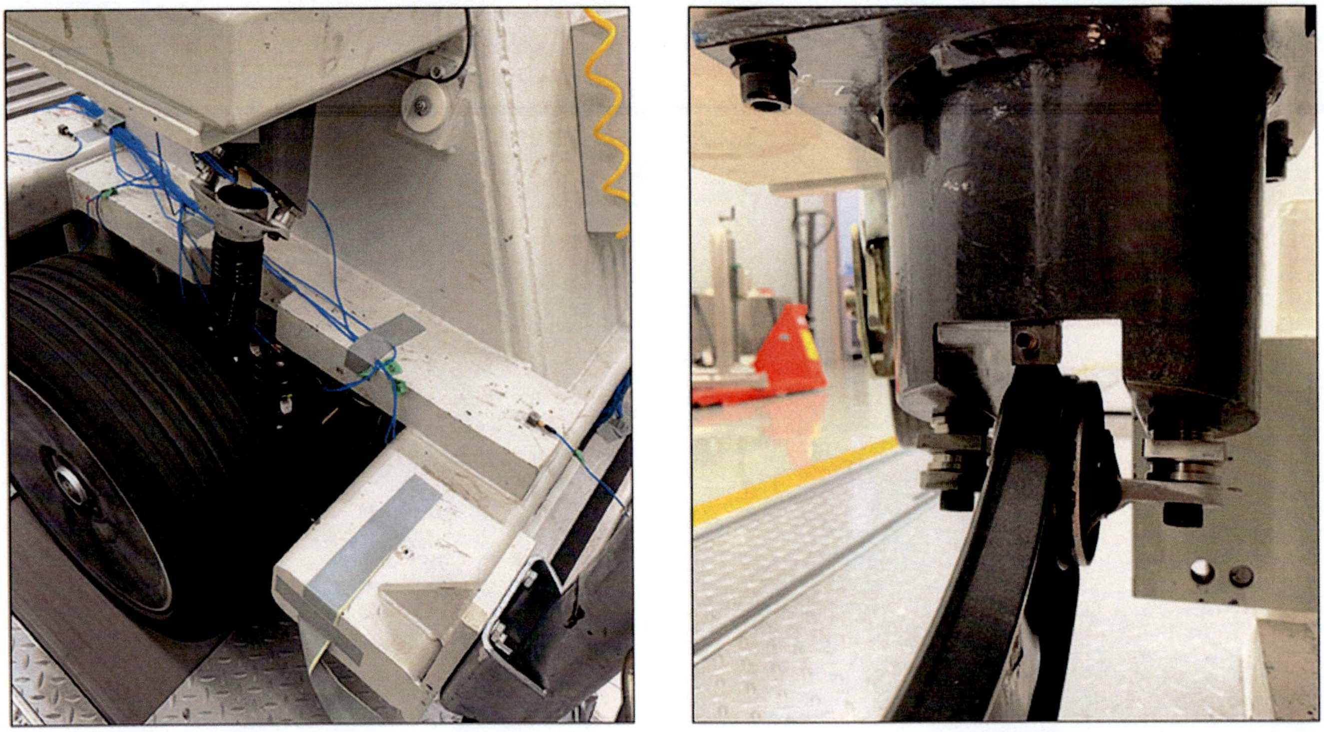

Fig. 3.6 Rigid test rig used for measuring blocked forces

While additional data could have been gathered for the dampers and springs, it was assumed that these would not be the critical load paths and was therefore out of scope of this work.

3.3.4 *Interface Forces: Derived from Blocked Forces*

Finally, it is also possible to convert blocked forces ($\mathbf{f}_2^{bl}$) to interface forces ($\mathbf{g}_2^{B}$) with:

$$\mathbf{g}_2 = \left(\mathbf{Y}_{42}^{B}\right)^{+} \mathbf{Y}_{42}^{AB} \mathbf{f}_2^{bl} \tag{3.4}$$

where $\mathbf{Y}_{42}^{B}$ and $\mathbf{Y}_{42}^{AB}$ are FRF matrices of the trimmed body and full vehicle, respectively. As mentioned in the Matrix Inverse section above, these FRF matrices have inputs at the interfaces and responses at indicator sensors around the interfaces. If the full-vehicle FRFs cannot be measured, this equation could be rewritten and simplified using a substructured approach [7] for $\mathbf{Y}_{42}^{AB}$; in this work, the full-vehicle FRFs were measured.

An example setup for the trimmed body FRF measurement was shown above in Fig. 3.3. The full-vehicle FRF measurement is quite similar, except impacts are made on the installed suspension rather than the transmission simulators used on the trimmed body.

The blocked forces ($\mathbf{f}_2^{bl}$) in this approach can be obtained in many ways, as detailed in [1]. In this study, the blocked forces were obtained by directly measuring them on a rigid test rig, as shown in Fig. 3.6. The rigid test rig has been used and described in previous studies [8, 9].

3.3.5 *Trimmed Body Measurements*

Once interface forces ($\mathbf{g}_2$) are known, they are used with trimmed body FRFs/NTFs ($\mathbf{Y}_{32}^{B}$) to predict noise levels at the driver's ear ($\mathbf{u}_3$). The FRFs/NTFs must have inputs at the same locations as the interface forces and responses at the locations

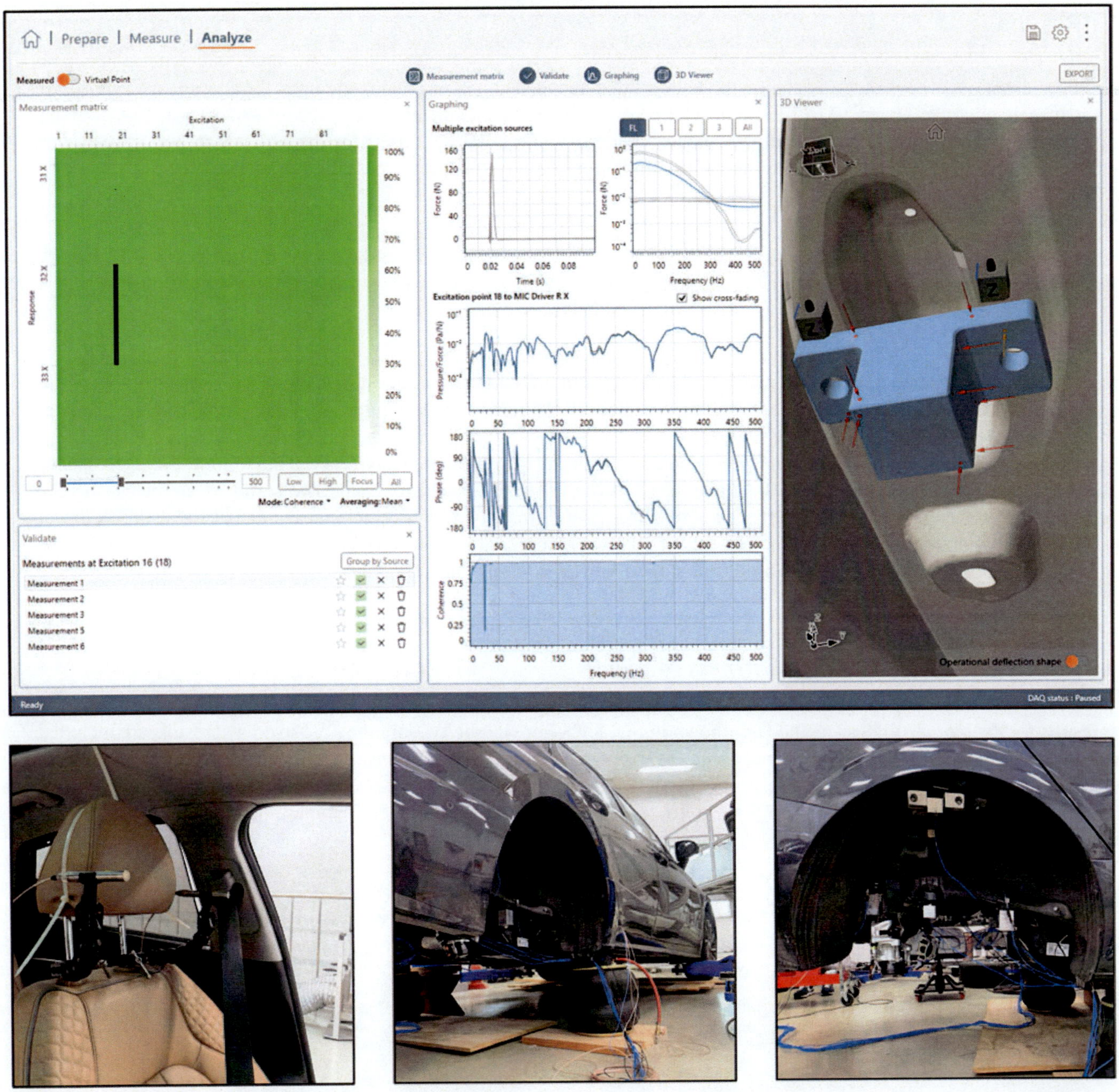

Fig. 3.7 Noise transfer function (NTF) measurement of the trimmed body in DIRAC. (*top*) Two impact hammers were used to properly excite the full frequency range of 0–500 Hz; their data (gray curves) is merged directly in the estimation of the transfer functions (blue curve). (*bottom*) Various photos showing the test setup of the trimmed body test

of interest (in this case, microphones at the driver's ear). The responses are then predicted with:

$$\mathbf{u}_3 = \mathbf{Y}^{\mathrm{B}}_{32}\mathbf{g}_2 \tag{3.5}$$

These trimmed body measurements are performed with ease in DIRAC, as shown in Fig. 3.7. Note that two impact hammers are used to excite the low and mid-frequencies; DIRAC filters and merges the data appropriately to gain a highly coherent transfer function with good signal-to-noise ratio for the entire frequency range of interest.

3.4 Results

Let us now discuss the application of the methods described above. The operational cycle used for all cases is a run-up of the rear wheels on a dyno from 20 to 120 km/h. The recorded time data is converted to 1-second blocks with 50% overlap and Hann-windowed before computed into frequency spectra. While DIRAC is used for the acquisition and virtual point transformation of all FRFs, all processing and TPA simulation is performed using SOURCE and the VIBES Toolbox for MATLAB.

3.4.1 Interface Forces: Matrix Inverse Method

Let us first look at the application of the interface forces obtained using a matrix inverse. Figure 3.8 shows the simulation at the left driver ear compared to the direct measurement as a Campbell diagram, at the same color scale. It is clear to see that all relevant noise phenomena in the car are well represented by the matrix inverse interface forces applied to the trimmed body noise transfer functions.

Another way of comparing the classical TPA prediction with the measurement is by looking at the time-averaged spectra of driver ear sound pressure levels in dB(A) over the full run-up from 20 to 120 km/h. The result of the matrix inverse interface forces is shown in Fig. 3.9a in green, where comparison is made with the vehicle measurement for a similar run-up (black). The horizontal grid lines represent intervals of 10 dB(A). There is clearly a very high agreement, which demonstrates that the selection of sensors and impacts used for the matrix inverse were appropriate.

3.4.2 Interface Forces: Mount Stiffness Method

The results for the mount stiffness method are shown in Fig. 3.9b. For simplicity, only the time-averaged results are included. The results from this method show similar accuracy to the matrix inverse method. This is particularly interesting because, as mentioned above, interface forces were only included at the six interfaces that contain resilient mounts (subframe connections and trailing arms). These results confirm that the forces at the unincluded interfaces (damper and spring) contribute very minimally to the overall noise levels at the driver's ear.

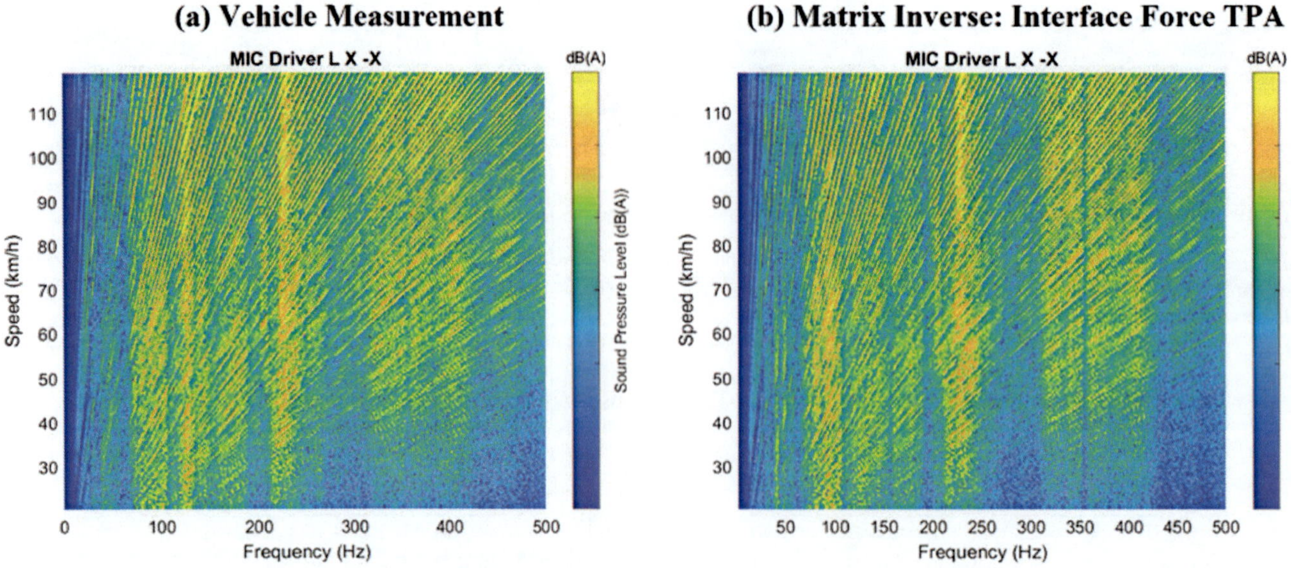

Fig. 3.8 Full spectra of the (**a**) validation measurement and (**b**) interface force TPA prediction using matrix inverse

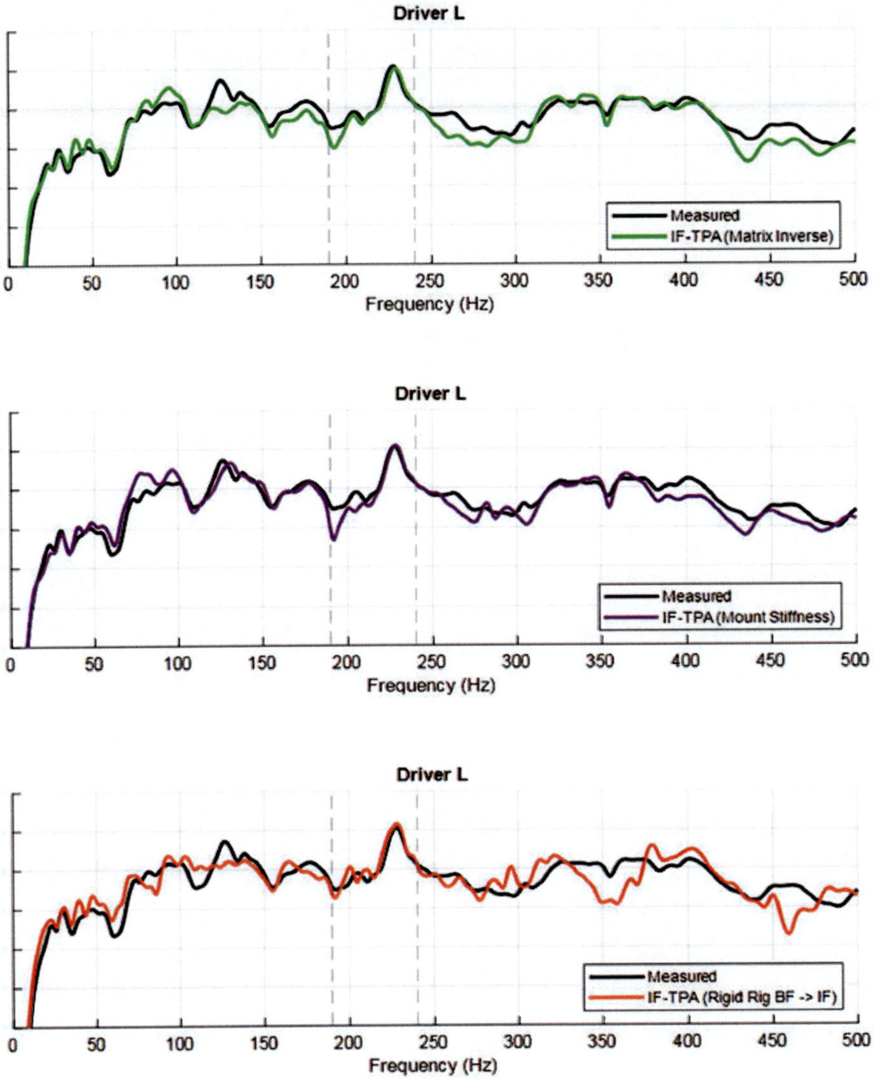

(a) Interface forces: matrix inverse

(b) Interface forces: mount stiffness method

(c) Interface forces: derived from measured blocked forces on test rig

Fig. 3.9 Result of applying interface forces to trimmed body FRFs to predict tire noise at the driver's ear, in time-averaged dB(A) spectra

3.4.3 Interface Forces: Derived from Blocked Forces

Finally, we present a hybrid approach for interface forces calculated during the run-up. This method is perhaps the most interesting/useful in terms of the workflow used. With interface forces derived from blocked forces, the engineer has all the tools in his toolbox to perform just about any transfer path analysis imaginable. The prediction is shown in Fig. 3.9c for interface forces derived from blocked forces; again only the time-averaged results are shown. Again, the comparison between validation measurement and prediction is quite good; most critically, there is a strong match in the frequency band dominated by a tire cavity mode (190–240 Hz).

3.5 Conclusions

In this paper, we showed how interface forces are calculated within the TPA framework using three different methods: a matrix inverse method, a mount stiffness method, and with interface forces derived from blocked forces. Their quality was confirmed by evaluating their capability to predict the road noise at the driver's ear. All methods strongly relied on the virtual point technology implemented in DIRAC to determine the interface forces and to ensure compatibility at the interfaces of different experimental models. Within the three methods, we consider the interface forces derived from blocked forces to add most value for the engineer during the development process. This approach allows for classic TPA investigations as well as more advanced analyses using frequency-based substructuring to study modifications on the receiving structure.

Acknowledgments We like to acknowledge NVH Research Lab of Hyundai Motor Company for funding this research. We like to thank Müller-BBM VibroAkustik Systeme Korea for their support in the measurement campaigns and coordination.

References

1. van der Seijs, M.V., Harvie, J.M., Song, D.P.: Road noise NVH: embedding suspension test benches in NVH design using Virtual Points and the TPA framework. In: Proceedings of the thirty-ninth international modal analysis conference (2021)
2. van der Seijs, M.V., Rixen, D.J., de Klerk, D.: General framework for transfer path analysis: history, theory and classification of techniques. Mech. Syst. Signal Process. **68**, 217–244 (2016)
3. Barten, E.: A complex power approach to characterise joints in experimental dynamic substructuring. Master's thesis, Delft University of Technology (2013)
4. van der Seijs, M.V., van den Bosch, D.D., Rixen, D.J., de Klerk, D.: An improved methodology for the virtual point transformation of measured frequency response functions in dynamic substructuring. In: COMPDYN 2013: 4th ECCOMAS thematic conference on computational methods in structural dynamics and earthquake engineering, Kos, Greece (2013)
5. Mayes, R.L.: Tutorial on experimental dynamic substructuring using the transmission simulator method. In: Proceedings of the thirtieth international modal analysis conference (2012)
6. Häußler, M., Klaassen, S.W.B., Rixen, D.J.: Experimental twelve degree of freedom rubber isolator models for use in substructuring assemblies. J. Sound Vib. **474**, 115253 (2020)
7. van der Seijs, M.V.: Experimental dynamic substructuring: analysis and design strategies for vehicle development. PhD dissertation, Delft University of Technology (2016)
8. Kang, Y.J., Song, D.P., Ih, K.D.: Estimation of body input force transmission change due to parts' modification using the impedance method under rolling excitation. Proc. Inst. Mech. Eng. D. J. Automob. Eng. **233**(2), 363–377 (2019)
9. Song, D.P., Min, D., Kang, Y.J., Cho, M., Kim, H.G., Ih, K.D.: A methodology for evaluating the structure-borne road noise prior to a prototype vehicle using direct force measured on a suspension rig. Noise Contr. Eng. J. **64**(3), 295–304 (2016)

Chapter 4
Real-Time Hybrid Substructuring for Shock Applications Considering Effective Actuator Control

Christina Insam, Michael J. Harris, Matthew R. Stevens, and Richard E. Christenson

Abstract Shock describes a rapid change in loading conditions and occurs in many mechanical, aerospace, and civil engineering systems. The shock response of these systems is of critical importance in their design and must therefore be studied. While experimental investigation of shock response offers accurate results, this approach is costly and requires highly specialized and unique facilities. In contrast, numerical investigation of shock events can be an effective alternative; however, modeling the systems accurately can be challenging. In this paper, the application of Real-Time Hybrid Substructuring (RTHS) to study the system response to a shock event is proposed. RTHS is a cyber-physical testing method, combining both experimental and numerical testing. The RTHS approach is intended to fully incorporate the dynamic interaction between the structure and the excitation source and realistically capture all dynamic phenomena. In this preliminary study of an RTHS shock test, the impact of a swinging pendulum on a mass–spring–damper system is investigated. This highly dynamic event requires precise actuator control and dynamics compensation. This work makes use of a model-based feedforward compensator, namely a minimum phase inverse compensator. To reduce any remaining frequency-dependent time delay or magnitude tracking errors, this compensator is combined with a P-type Iterative Learning Controller. The interaction force profile is studied for varying eigenfrequencies and mass ratios of the impacted mass–spring–damper system. The tests are able to replicate the free vibration response of the system accurately. Despite a good learning performance of the Iterative Learning Control, there are still tracking errors in the initial impact phase. Future work will look to improve actuator control and performance.

Keywords Real-time hybrid substructuring · Shock · Actuator compensation · Iterative learning control · Model-based feedforward compensator

4.1 Introduction

Shock response is of importance in the design of mechanical, aerospace, and civil engineering systems. While the definition of the term shock has not been formalized, it is ubiquitously synonymous with a distinctly rapid change in loading conditions [11]. Examples of shock loading conditions include blast events for structural, space, and maritime systems [8, 9, 28], fluid interactions such as sound and wind [11], as well as impact events such as ship impacts with bridge piers [1]. Depending on the bandwidth of the excitation, the response of a system to shock loading can be divided into two regimes: near-field response and far-field response. The near-field response is characterized by high-frequency oscillations (upwards of 10 kHz) and tends to damp out quickly. The far-field response, in turn, is dominated by lower-frequency oscillations, which tend to define the response of the system on more of a global scale. Far-field oscillations typically contain energy at frequencies corresponding to the structural modes of vibration [11].

Prediction of the shock response of a system can be performed either experimentally or numerically. Experimental shock response testing for scenarios such as impact events is typically conducted using a variation of a drop machine. In a drop machine, a mass falls under the full or partial influence of gravity and subsequently impacts a test specimen, whose

C. Insam
Technical University of Munich, School of Engineering and Design, Department of Mechanical Engineering, Chair of Applied Mechanics, Munich, Germany
e-mail: christina.insam@tum.de

M. J. Harris · M. R. Stevens · R. E. Christenson (✉)
University of Connecticut, Department of Civil and Environmental Engineering, Storrs, CT, USA

M. Allen et al. (eds.), *Dynamic Substructures, Volume 4*, Conference Proceedings of the Society for Experimental Mechanics Series,
https://doi.org/10.1007/978-3-031-04094-8_4

response is then analyzed [11]. A governing assumption associated with drop machine testing is that the specimen and its dynamics have little effect on the force placed on the specimen. This assumption is generally valid when the impacting mass is small in comparison to the dynamics of the impact test rig and the impacting mass can be assumed to be contained within a bandwidth far above those of the test specimen [11]. If this is not the case, the interactions of dynamics within the test specimen and impacting mass will alter the force profile exerted on the test specimen. Hence, experimental testing is effective in capturing the near-field response; however, the dynamic interaction of the excitation loading and global structural behavior are neglected. Numerical simulation can also be used to predict the shock response of a system. In order to do so, the excitation loading must be characterized and applied to a numerical representation of the structure [11]. Subsequently, numerical analysis can be accomplished by imparting the excitation loading on the entire test specimen as a multi-degree of freedom (MDOF) system or by exciting an ensemble of single degree of freedom (SDOF) systems with varying resonant frequencies. By using the latter method, it is possible to obtain the shock response spectrum for a given excitation. Even though numerical simulation of shock events can be a quite effective alternative to experimental testing, it may be challenging to accurately model the near-field response behavior with potential uncertainties associated with high nonlinearities and the challenge of capturing the excitation loading and dynamic interaction with the far-field and global response of the structure. In this paper, Real-Time Hybrid Substructuring (RTHS) is demonstrated to study the shock response of a structure. RTHS, also called Real-Time Hybrid Simulation or Real-Time Dynamic Substructuring, combines experimental testing and numerical simulation. Early developments of RTHS include Horiuchi et al. [13, 14], Nakashima and Masaoka [21, 22] and Darby et al. [7]. With RTHS, a structural dynamic system can be partitioned into separate experimental and numerical components or substructures and interfaced together as a cyber-physical system. The substructures that are well understood are simulated in real-time using analytical or numerical models, while the substructures of particular interest, highly complex, or nonlinear are tested experimentally using physical specimens. In an RTHS test, the experimental and numerical substructures communicate together in real-time at a fixed sampling time of δt. This is achieved by transferring motion and force signals through a feedback loop using controlled actuation, sensing, and a Digital Signal Processor (DSP). The signal flow of this basic RTHS setup is visualized in Fig. 4.1. Advances in numerical computing power, digital signal processing, and high-speed actuation enable testing of structures with higher frequencies and/or higher frequency excitations [29]. By using RTHS for shock testing, the near- and the far-field behavior of the system and the dynamic interaction with the excitation loading can be analyzed. The near-field response such as local high-frequency responses and contact behavior will be captured through the physical substructure and the far-field dynamics will be simulated within the numerical substructure.

In this work, the dynamic interaction of a pendulum system (experimental substructure) impacting a mass–spring–damper (MKC, numerical substructure) system is investigated using RTHS. An electrodynamic shaker is used for actuation. A key challenge in the realization of an RTHS test for shock events is the required accurate actuator tracking to replicate this highly dynamic event. For this purpose, the combined use of a minimum phase inverse compensator (MPIC) [3] and P-type Iterative Learning Control (ILC) [15, 16] is proposed and the efficacy is investigated. In addition, the force profiles are analyzed for varying mass values and eigenfrequencies of the MKC system.

The rest of the paper is structured as follows: Sect. 4.2 introduces the MPIC and ILC used to compensate the actuator dynamics. In Sect. 4.3, the dynamical system, the experimental setup, and the parameters used for the experimental tests are

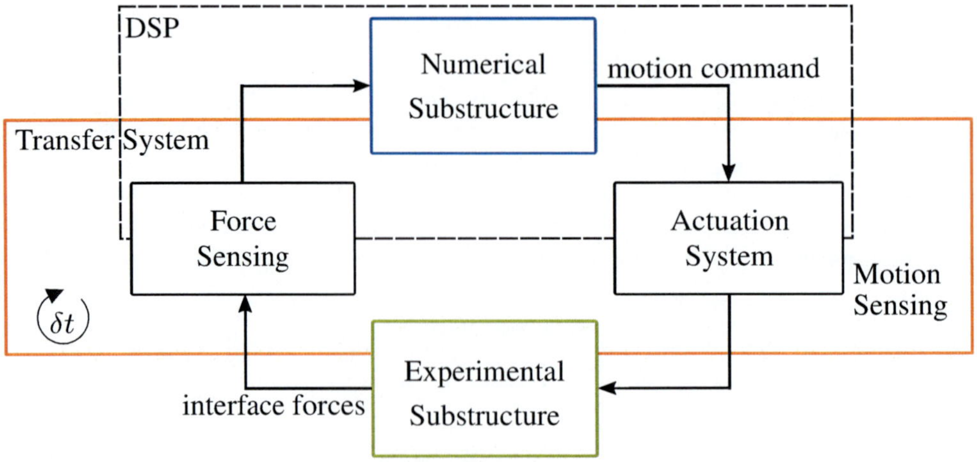

Fig. 4.1 RTHS forms a closed-loop system, where the numerical and the experimental substructure exchange motion and force information in real-time with each other. The transfer system comprises the actuation system, sensors, and a DSP

described. The results of the tests are provided in Sect. 4.4. Section 4.5 summarizes the findings and concludes the paper by presenting the research challenges associated with RTHS testing of shock events.

4.2 Introduction to Compensators in Real-Time Hybrid Substructuring

Since the RTHS loop forms a closed-loop system, the fidelity of the test results and also the test stability highly depend on the tracking performance of the actuator. Hence, the compensation of the actuator dynamics and any delays in the loop has been a vast research topic in the field of RTHS over the last two decades. The first delay compensation scheme was proposed by Horiuchi et al. [13, 14] and makes use of a polynomial extrapolation of the actuator command. After that, more advanced techniques from the realm of control theory have been proposed for the application in RTHS, see i.a. [2, 10, 20]. In this work, the combined use of MPIC and ILC is proposed. Their function as well as their advantages and shortcomings are presented in the following.

4.2.1 Minimum Phase Inverse Compensator

Model-based feedforward control compensates for the actuator dynamics by using an inverse model of the actuator $\hat{P}$. With an accurate actuator model P, its inverse can be feedforwarded to cancel poles and zeros of the actuator dynamics. The open-loop response of the compensated actuator would result in unity gain and zero-phase error. Model-based controllers have been used in the work by Carrion and Spencer [5, 6], Phillips and Spencer [26, 27] and Fermandois and Spencer [10]. These implementations differ in the method that is used to obtain the actuator model and the technique to invert the transfer function. Model-based feedforward control enables high tracking performance of the actuator over a large frequency bandwidth, but also offers several challenges:

- A system identification is required to model the actuator dynamics.
- The magnitude and phase properties of the actuator need to be accurately modeled in the frequency range of interest to achieve a high tracking accuracy.
- For this compensator design, the inverse of the identified actuator dynamics is required to be stable and causal [3]. If P is strictly proper, its inverse is improper and the transfer function should be padded by adding a low-pass filter [6]. In addition, any zeros of P with positive real parts become unstable poles when inverted [3].

A minimum phase inverse compensator (MPIC) [3] can be used to obtain a stable and causal compensator that does not require additional low-pass filtering. The underlying theory is that a stable and causal transfer function can be split into a minimum phase part and an all-pass transfer function. The minimum phase part, as well as its inverse, has the property of being causal and stable, which means that there are no right-hand side poles or zeros. While the minimum phase part of a transfer function captures the magnitude of the transfer function, the phase is not correctly captured. The phase delay information is stored in the all-pass part of the transfer function. In MPIC, only the minimum phase part is used for the compensator design, which is further denoted by P_{mp}. Using this approach, accurate magnitude tracking can be assumed at the cost of a remaining phase error. Botelho [3] states that an additional delay compensator can be used to compensate the remaining phase error using this approach.

This work makes use of the MPIC, since the implementation approach is straightforward and does not add any additional dynamics. However, for this study, MPIC as a standalone compensator resulted in unacceptable phase errors across the control bandwidth. Hence, an additional controller is required to correct for the errors of the MPIC.

4.2.2 P-Type Iterative Learning Control

Tracking performance of a model-based feedforward control (e.g., due to modeling errors) can be improved by combining a controller to ensure robustness. This is often a feedback controller, such as a simple linear proportional–integral controller or a linear–quadratic regulator [10]. In this work, Iterative Learning Control (ILC) is proposed to enhance the compensator's performance.

ILC is a feedforward control scheme that can be used whenever a manipulator performs the same motion repetitively [4, 23–25]. The underlying thought is that—if the controllers of the manipulator, the motion command, or the system initial state do not change—the same error occurs every time the robot performs this motion task. The idea in ILC is to learn from previous iterations and find a feedforward signal that minimizes this error. This is done by storing the error signal between the commanded and the achieved motion e_j (iteration j) throughout the motion execution. Then, the error signal is processed by a learning function L and used as a feedforward signal $f_{\mathrm{j}+1}$ the next time this motion is performed, i.e., in iteration $j+1$. So, even though ILC is a feedforward controller during one iteration, it can be considered a feedback controller in iteration-domain.[1] The huge advantage of ILC is that the error signal of the whole motion trajectory from the previous iteration $j-1$ is available. This enables ILC to have anticipating properties.

ILC has first been applied to RTHS in the work of Hochrainer and Puhwein [12]. Insam et al. studied the properties of ILC in RTHS in more detail in [15, 16]. The application of ILC to RTHS problems is straightforward and offers an improvement of the actuator tracking performance. Using ILC in RTHS requires that the RTHS test can be performed several times in a row. In [16], a convergence condition is derived, which enables to properly choose the parameters of the learning function L. In prior work [15, 16] as well as in this paper, a relatively simple implementation of ILC is used, namely P-type (proportional) ILC [4]. Here, the error signal is processed by a proportional gain β. The feedforward signal $f_{\mathrm{j}+1}$ for the next iteration $j+1$ writes

$$f_{\mathrm{j}+1} = Le_\mathrm{j} + f_\mathrm{j} = \beta e_\mathrm{j} + f_\mathrm{j}, \tag{4.1}$$

where the feedforward signal of the current iteration is f_j. To increase the robustness of the scheme, a robustness filter Q is added. This is often a zero-phase low-pass filter with cutoff frequency $f_{\mathrm{Q,cut}}$. Note that zero-phase filtering of the error signal is possible, because the error signal of the whole motion trajectory of the previous iteration is available. Using the robustness filter, the update equation of P-type ILC writes

$$f_{\mathrm{j}+1} = Q\left(Le_\mathrm{j} + fj\right). \tag{4.2}$$

The length of the error vector e_j is $\frac{T}{\delta t}$, where T is the duration of the motion task and δt is the fixed sampling time. To improve the anticipating behavior of ILC further and make the algorithm even anti-causal,[2] a time shift δ can be implemented. This means that, instead of $e_\mathrm{j}(t_\mathrm{k})$ with $t_\mathrm{k} \in (0, T)$, the error vector is shifted forward by δ samples. The shifted error signal $\tilde{e}_\mathrm{j} = e_\mathrm{j}(t_\mathrm{k} + \delta)$ is then used in Eq. (4.2).

In the application of RTHS to shock response, a quick and accurate actuator motion is crucial. Due to the promising properties of ILC, such as the anticipating behavior and the potential to overcome frequency-dependent delays [16], this control scheme is applied in this paper to complement MPIC.

4.3 Experimental Setup

This section explains the used RTHS setup as well as the compensator design and the parameters used in the conducted tests. An overview about the RTHS setup is given in Fig. 4.2 and the individual parts are explained in the following.

4.3.1 Dynamical System

In this study, a swinging pendulum impacting a mass–spring–damper (MKC) system is investigated using RTHS. One end of the MKC system is grounded. This system is suited to demonstrate RTHS shock testing, as it easily allows the dynamic properties of the MKC system to be varied. The interaction force as well as the free vibration response can be studied using this simple LTI system. The dynamic response of the system can be considered to consist of two phases: the interaction response followed by the free vibration response. Generalized to an MDOF structure, the interaction response is analogous

[1] For example, a proportional–integral controller is a feedback controller in time domain.

[2] An anti-causal system is a system, where the output depends on future inputs.

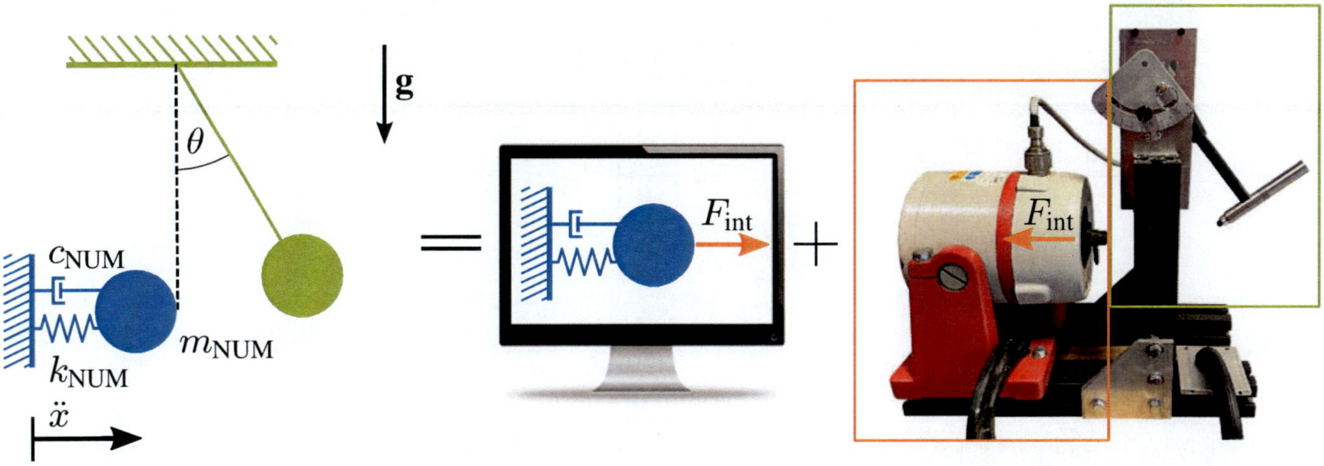

Fig. 4.2 In the RTHS test, a hammer (experimental part, in green) impacting an MKC system (numerical part, in blue) is investigated. The transfer system (in orange) comprises a controlled electrodynamic shaker and an impedance sensor to measure the interface forces and accelerations. Furthermore, a performance real-time target machine by Speedgoat is used for DSP to exchange the signals between the numerical part and the physical setup (not displayed in the figure)

to a near-field structural response in which local high-frequency dynamics and nonlinear deformations of the contact partners dictate the dynamic response. Furthermore, the free vibration response is analogous to a far-field response in which natural frequencies and mode shapes dictate the dynamic response.

The MKC system forms the numerical substructure, and the pendulum and the interaction forces are represented experimentally. At the beginning of the tests, the pendulum and the MKC are at system rest, with the pendulum having an initial deflection of θ_0. The MKC system consists of a mass m_{NUM}, a spring with spring constant k_{NUM}, and a damper with damping constant c_{NUM}. The equation of motion is

$$m_{\mathrm{NUM}}\ddot{x} + c_{\mathrm{NUM}}\dot{x} + k_{\mathrm{NUM}}(x - x_0) = F_{\mathrm{int}}, \tag{4.3}$$

with the resting spring length x_0. The interface force F_{int} is the only excitation of the MKC system.

In the proposed RTHS test, the dynamic interaction of the pendulum and the MKC system (far-field dynamics) will be fully captured, while the physical behavior of the shock loading (the pendulum) will be experimentally tested along with any near-field response at the impact zone.

4.3.2 Actuation System

An electrodynamic shaker (model LDS V406, Brüel and Kjær, Denmark) is used as the actuation system. Since the shock response of the dynamic system contains high frequencies and requires a large bandwidth, a shaker was used for the RTHS in lieu of commonly used servo-hydraulic actuators. The shaker command (in voltage) is sent to an amplifier (model XLS DriveCore 2 Series, Crown International, USA), which outputs a current proportional to the command voltage. Electrodynamic shakers are driven by Lorentz forces, which means that the shaker force is proportional to the input current. If the structure that is driven by the shaker is only a mass, the output acceleration is also proportional to the input current [17]. In order to use this proportionality and due to the lack of a displacement sensor, acceleration control is used in this RTHS setup (acceleration command $\ddot{x}_{\mathrm{C}}$). To improve the command response of the shaker, a combination of MPIC and ILC is employed in this work. The implemented control scheme is visualized in Fig. 4.3.

The implementation of MPIC requires the minimum phase part P_{mp} of the actuator model P. This comprises the amplifier and the shaker. For that purpose, a system identification with band-limited white noise (BLWN) as input signal was performed [19]. The measuring frequency was 25,600 Hz for a measurement period of 60 s . The measurement was acquired using a DataPhysics Abacus 906 signal analyzer (DataPhysics Corporation, USA) and post-processed using MATLAB. A Hanning window with 50% overlap was applied and the resulting measurement of the transfer function was calculated with a frequency resolution of 0.1 Hz. Using the measured transfer function, a function fit with a model of five zeros and five poles was performed. Then, the minimum phase part of the transfer function was taken and inverted to get the MPIC. The resulting

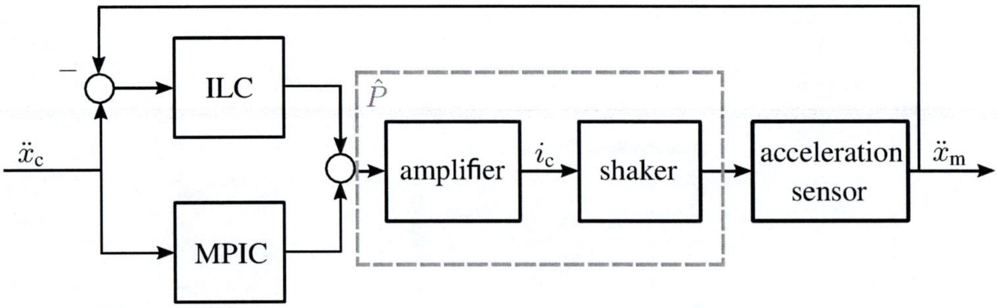

Fig. 4.3 The shaker is controlled by an MPIC and ILC. To make use of the proportional relationship between shaker input current i_c and shaker acceleration ($\ddot{x}_m$), acceleration control is used

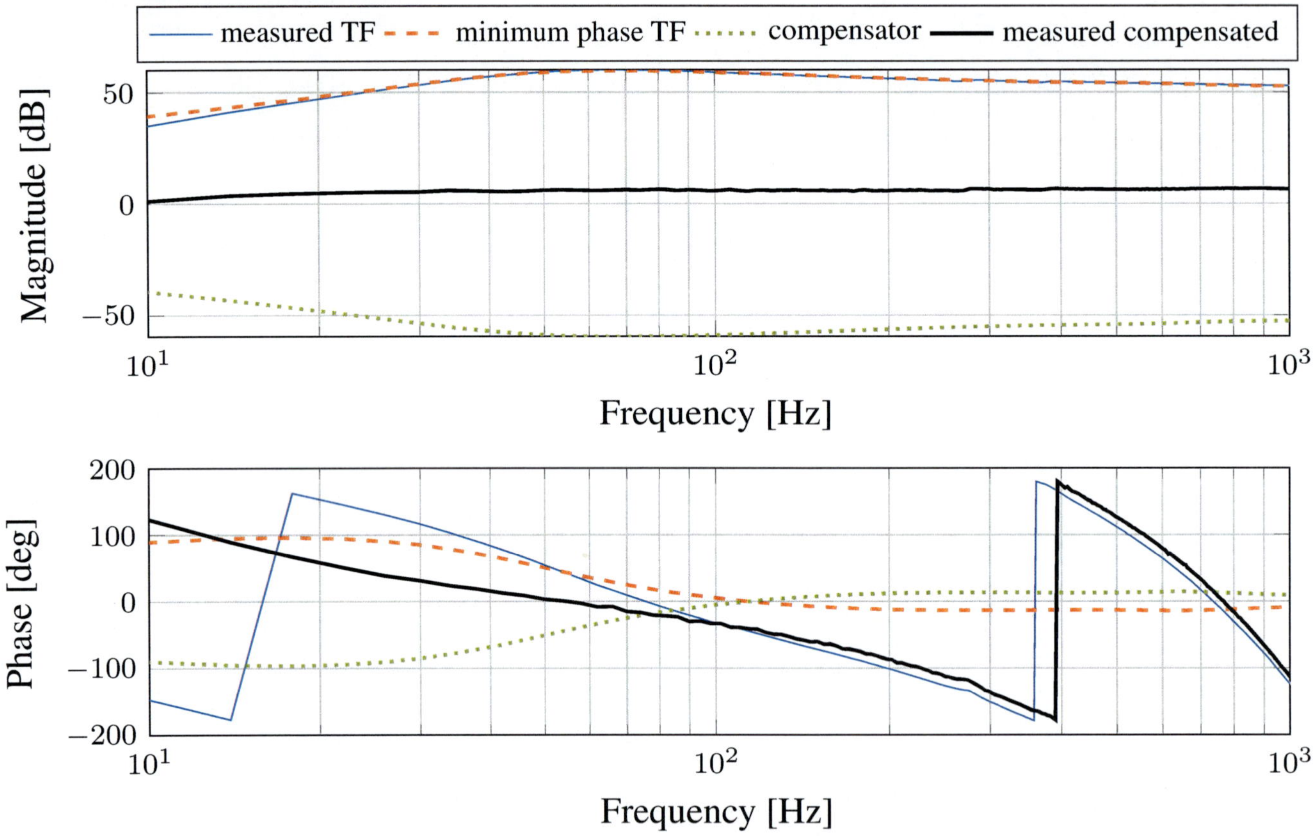

Fig. 4.4 The bode plot of the measured transfer function (TF) of the electrodynamic shaker $\hat{P}$, the minimum phase approximation P, the minimum phase inverse compensator transfer function G_{MPIC}, and the measured compensated transfer function $G_{\text{MPIC}}\hat{P}$

transfer function of the MPIC writes

$$G_{\text{MPIC}} = \frac{1}{P_{\text{mp}}} = \frac{s^5 + 3.64 \cdot 10^3 s^4 + 1.88 \cdot 10^7 s^3 + 1.71 \cdot 10^{10} s^2 + 5.42 \cdot 10^{12} s + 8.96 \cdot 10^{14}}{4.17 \cdot 10^2 s^5 + 1.89 \cdot 10^6 s^4 + 9.1 \cdot 10^9 s^3 + 1.23 \cdot 10^{13} s^2 + 1.27 \cdot 10^{15} s + 1.75 \cdot 10^{16}}. \tag{4.4}$$

The Bode plots of the measured and the approximated transfer function of the shaker as well as the transfer function of the MPIC are shown in Fig. 4.4 in a frequency range up to 1000 Hz. Furthermore, the measured compensated transfer function $G_{\text{MPIC}}\hat{P}$ is shown. As can be seen, the magnitude is close to unity and the phase is almost linear corresponding to a time delay of ≈0.002 s.

4.3.3 Experimental Substructure

The experimental part is formed by a modal hammer (ICP Impact Hammer 086C03, PCB Piezotronics, USA), which is mounted with a pivot joint. The swinging hammer forms a pendulum system. The initial deflection θ_0 of the hammer can be adjusted and fixed with a locking mechanism (cf. Fig. 4.2). The pendulum can be released easily by pulling the lock. The pendulum hits the shaker with its rubber tip when $\theta = 0°$. In preliminary experiments, the eigenfrequency of the swinging pendulum (mass 0.16 kg, length 0.18 m) was identified as 7.285 rad/s with a damping ratio of 1.5%.

4.3.4 Sensors

To close the RTHS loop, the interface forces F_{int} acting between the pendulum and the MKC system need to be measured and fed back to the numerical simulation. Furthermore, the achieved shaker acceleration $\ddot{x}_{\text{m}}$ needs to be measured for the application of ILC (cf. Fig. 4.3). To measure both the force and acceleration, a mechanical impedance sensor (model TLD288D01, PCB Piezotronics, USA) is mounted at the tip of the shaker.

When the shaker oscillates, the impedance sensor outputs forces. This is due to the inertial forces of the impedance sensor itself. In common applications of impedance sensors, this inertia force can be neglected compared to the force range of interest. Here, these inertial forces need to be compensated because the inertia forces of the impedance sensor are not significantly smaller than the interface forces of the impact. To identify the portion of the sensor mass that creates these inertial forces, the transfer function from measured acceleration to measured force was identified using the aforementioned BLWN test. The transfer function is shown in Fig. 4.5a. As can be seen, the magnitude is constant in the whole frequency range. Furthermore, the direction of the force is opposite to the acceleration of the impedance sensor. This implies that the mass that needs to be compensated is $m_{\text{comp}} = 8.53 \cdot 10^{-3}$ kg (total sensor mass $19.2 \cdot 10^{-3}$ kg). Hence, the measured interface forces F_{int} that are used in the numerical simulation need to be corrected by $m_{\text{comp}}\ddot{x}_{\text{m}}$.

To verify this implementation, the interface forces of an RTHS test of the described dynamical system are shown in Fig. 4.5b. The impact duration, i.e., the interaction between the hammer and the impedance sensor is ≈ 0.012 s. Since there is no interaction after the impact, the measured interface forces should be zero for $t > 0.012$ s. As can be seen in the figure, this is the case for the compensated interface forces, while the uncompensated output signal of the impedance sensor shows the oscillations of the shaker. The modal hammer provides an additional force measurement. The force signal by the hammer

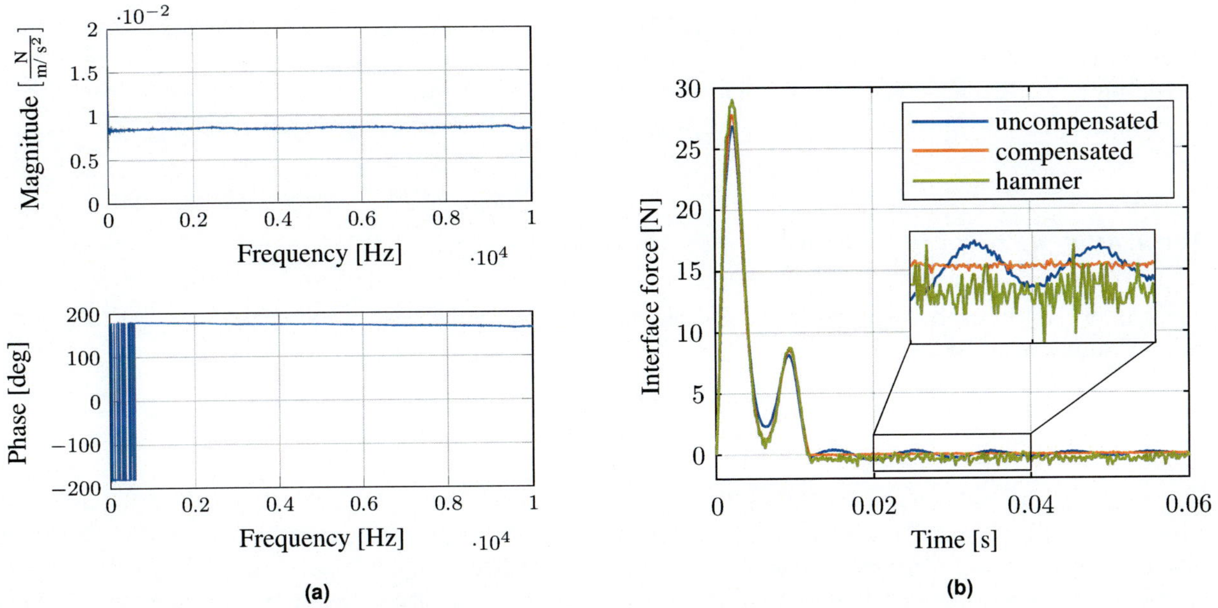

Fig. 4.5 Parts of the impedance sensor create inertial forces when the shaker oscillates. These forces need to be compensated in the intended RTHS test. (**a**) The bode plot from acceleration measurement to force measurement when the shaker was excited with BLWN. (**b**) The (un-)compensated interface forces as well as the forces by the impulse hammer in an RTHS test with impact

Table 4.1 Summary of the system parameters and parameters of the ILC implementation used in the RTHS tests

Test	Eigenfreq.	Mass ratio m_{ratio}	ILC parameters P gain β	Cutoff freq. $f_{Q,cut}$	Time shift δ	Test duration T
1	50 Hz	3	0.01 V/m/s²	150 Hz	0	0.5 s
2	100 Hz	3	0.005 V/m/s²	200 Hz	0	0.25 s
3	100 Hz	1	0.005 V/m/s²	200 Hz	3	0.25 s

is also displayed in Fig. 4.5b. In the following RTHS tests, the compensated force signal of the impedance sensor was used rather than the hammer signal. This is because, in a general RTHS setup of a shock test, the hammer signal would not be available. Furthermore, the measurement range of the hammer is up to 2000 N and therefore the results of the impedance sensor can be assumed to be more accurate in this test with maximum 30 N.

4.3.5 *Parameters*

The real-time application is implemented using MATLAB/Simulink (version R2021a, The MathWorks Inc., USA) and is run on a performance real-time target machine by Speedgoat. The RTHS tests were conducted with a fixed sampling frequency of $\frac{1}{\delta t} = 20\,\text{kHz}$ and the Dormand-Prince (ode 8) time integration scheme was selected. Three different RTHS tests were performed with varying eigenfrequencies of the MKC system and varying mass ratios between the numerical mass and the experimental mass ($m_{\text{EXP}} = 0.16\,\text{kg}$), i.e. $m_{\text{ratio}} = \frac{m_{\text{NUM}}}{m_{\text{EXP}}}$. The stiffness and damping parameter of the MKC system were tuned such that the damping ratio was 5%. In all tests, $\theta_0 = 15°$ was used as the starting angle of the pendulum. The shaker was controlled using a combined use of MPIC and P-type ILC, as described in Sect. 4.2, with the transfer function of the MPIC as given in Eq. (4.4). The parameters for the implementation of the P-type ILC were found using the convergence condition from [16], and a first order Butterworth filter was implemented as a robustness filter. The parameters of the three tests are summarized in Table 4.1.

4.4 Results and Discussion

The main goal of this work was to investigate whether RTHS tests of shock phenomena are feasible and to determine the major challenges. This implies that the interaction force as well as the free vibration behavior should be replicated by the RTHS test. For this purpose, tests with the RTHS setup as described in Sect. 4.3 were conducted. First, convergence of ILC needs to be verified. After that, the results of RTHS tests with different MKC systems will be presented.

4.4.1 *Verification of ILC Convergence*

ILC converges if an error measure decreases in subsequent iterations. Here, the relative root-mean-square (RMS) error is used as error measure, which is defined as [18]

$$e_{\text{rel}} = \frac{RMS(\ddot{x}_{\text{c}} - \ddot{x}_{\text{m}})}{RMS(\ddot{x}_{\text{c}})}. \tag{4.5}$$

The results are representatively shown in Fig. 4.6 for Test 2, cf. Table 4.1, and they look comparable for Test 1 and 3. As can be seen in the figure, the ILC scheme is able to reduce the error to about 40% of the initial error. In this test, ILC was not able to reduce the tracking error further for more iterations, i.e., ILC has already converged in iteration 8. The acceleration error $\ddot{x}_{\text{c}} - \ddot{x}_{\text{m}}$ is visualized in Fig. 4.7a for different iterations. This figure reveals that the shaker, when struck by the pendulum, tends to move at a higher acceleration than required. This shaker recoil cannot be countered by effective compensation due to a delay of 1.5 ms of the amplifier. For the remaining impact duration (> 1.5 ms), ILC manages to reduce the acceleration error.

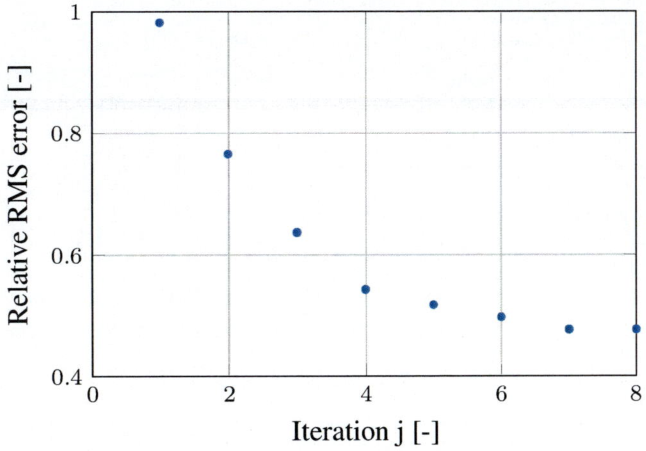

Fig. 4.6 The relative RMS error is reduced by the ILC to about 40% of the initial error within eight iterations

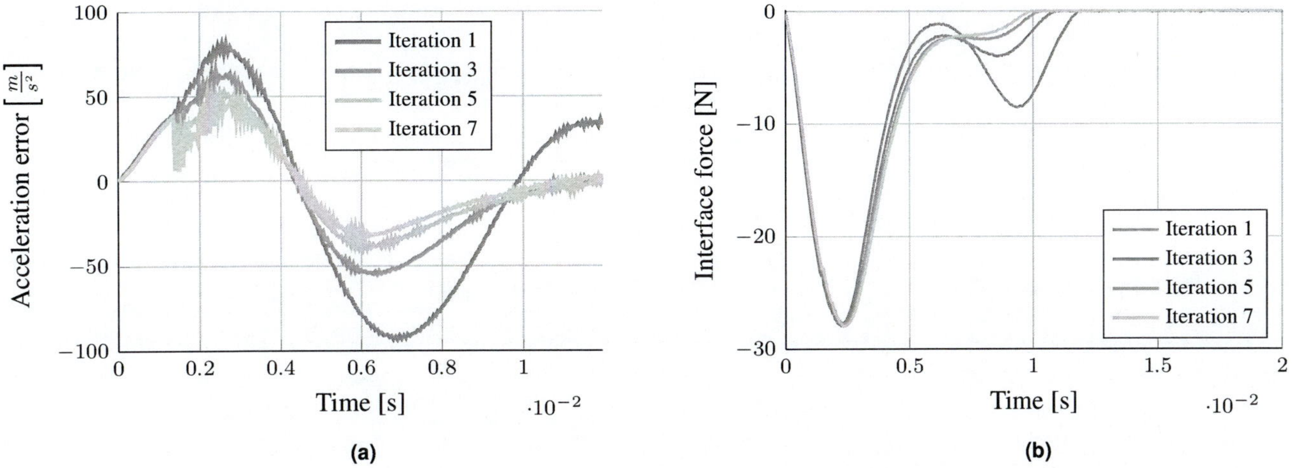

Fig. 4.7 In the initial impact phase, ILC is not able to reduce the acceleration error. After that, ILC leads to an improved tracking performance. Due to the improved tracking performance, an altered interaction force profile is observed. The results are representatively shown for the system parameters of Test 2 and show a similar behavior for Test 1 and 3. (**a**) The acceleration error $\ddot{x}_c - \ddot{x}_m$ during the impact in different iterations. (**b**) The interface force in different iterations

Figure 4.7b displays the interaction force profile for different iterations. The figure shows that the improved tracking performance leads to an altered force profile, which substantially differs in the formation of the second hump of the force profile. The second force hump decreases, which implies that the true interaction force profile for this MKC system (eigenfrequency and mass ratio) exhibits only one force peak. The first force hump is similar in all tests due to the shaker recoil. The shape of the force hump and the peak value would change if the amplifier delay was shorter or if the time instant of impact could be predicted and used to control the shaker.

To see more clearly how ILC changes the tracking behavior, the time domain data of the commanded as well as the measured acceleration are shown in Fig. 4.8. In the first iteration (only MPIC, no ILC feedforward), a large overshoot of the real acceleration compared to the commanded acceleration can be seen during the impulse. This overshooting behavior can substantially be reduced by ILC learning, cf. Fig. 4.8b. Nevertheless, the overshoot is still significant, which is in accordance with the results in Fig. 4.7. During the free vibration response of the MKC system,[3] a phase delay can be seen in the first iteration (Fig. 4.8a). This is due to the phase error of the minimum phase model used for the MPIC implementation. ILC manages to reduce the phase error and achieve almost perfect magnitude and phase tracking, cf. Fig. 4.8b.

[3] The free vibration response is technically not an RTHS test, because the interface force is zero.

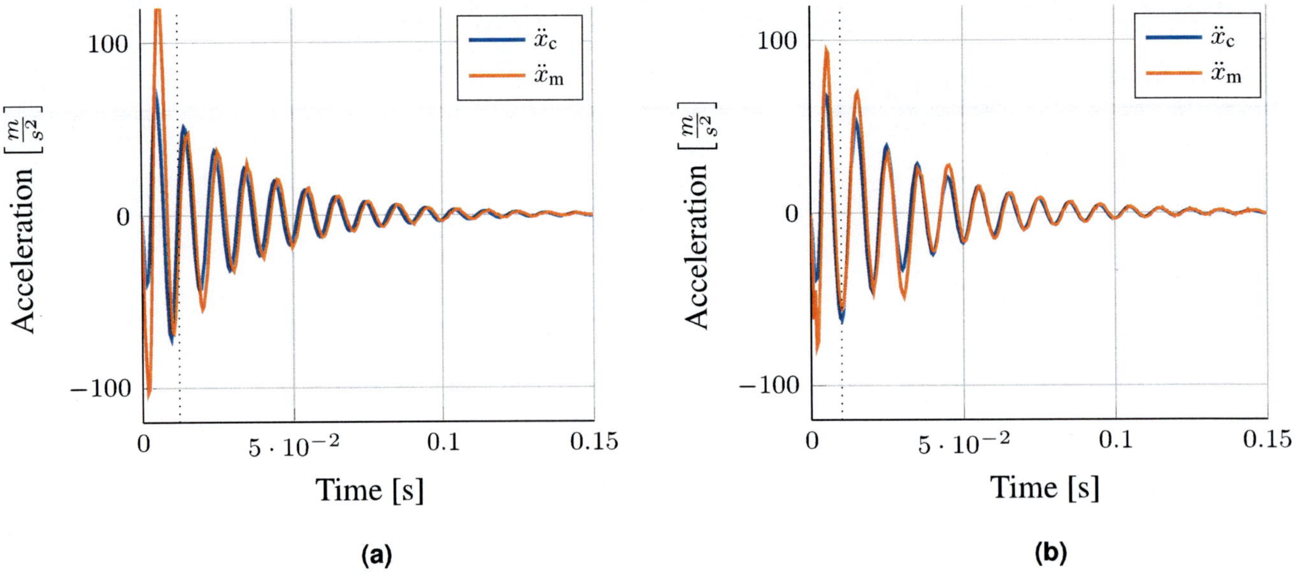

Fig. 4.8 The figures show the commanded acceleration $\ddot{x}_c$ and the measured acceleration $\ddot{x}_m$ of the shaker tip in Test 2. The vertical dotted line marks the end of the impulse interaction. (**a**) Results in the first iteration. (**b**) Results in the seventh iteration

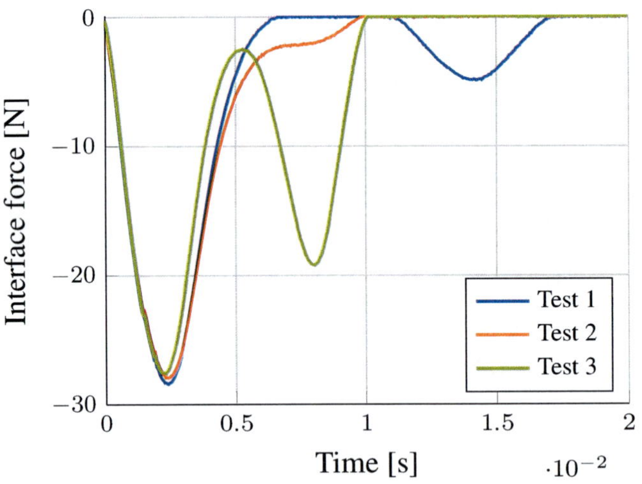

Fig. 4.9 The interface forces during the RTHS test differ for MKC systems with varying eigenfrequencies or mass ratio compared to the hammer mass (cf. Table 4.1). Test 1 with 50 Hz and $m_{\text{ratio}} = 3$, Test 2 with 100 Hz and $m_{\text{ratio}} = 3$ and Test 3 with 100 Hz and $m_{\text{ratio}} = 1$

4.4.2 Variation of the System Parameters

Figure 4.9 displays the interface force profiles for the RTHS tests of the three different MKC systems (see Table 4.1). In these tests, ILC has already converged (iteration 5 for Test 1 and iteration 7 for Test 2 and 3). The general trend is that the impact duration gets shorter for a higher eigenfrequency of the MKC system (Test 1 vs. Test 2 and 3).

The same impulse has to be transferred between the hammer and the MKC system in Test 1 and Test 2. This is because they have the same mass ratio. Since the impulse duration in Test 2 is shorter, a higher peak force would be expected but this is not observed—in fact, all three tests have almost identical peak forces. The reason for this is the shaker recoil, which cannot be countered by ILC learning and will be considered in future work.

The mass ratio is different for Test 2 and Test 3. As can be seen in Fig. 4.9, this considerably influences the shape of the interaction force. While Test 2 (mass ratio 3) develops only one force hump, two force peaks are developed in Test 3 (mass ratio 1).

The results imply that the RTHS test is in general able to replicate the interaction during the impulse duration. However, currently there is no reference solution available where the RTHS results can be validated against. A reference solution would be a fully numerical or a fully experimental result of the swinging hammer impacting the MKC system. In particular, modeling the impulse duration accurately to obtain the force profile is complex (e.g., contact between rubber tip of the hammer and impedance sensor). Furthermore, the shaker recoil during the hammer hit remains a challenge.

4.5 Conclusion

This work introduced shock testing using Real-Time Hybrid Substructuring (RTHS). For this purpose, an RTHS test of a pendulum impacting a mass–spring–damper system was conducted. An electrodynamic shaker was used as an actuator. An impulse test involves high dynamic bandwidth, as such, advanced actuator dynamics compensation is required. In this work, the combined use of a feedforward model-based compensator, specifically the minimum phase inverse compensator (MPIC), and P-type Iterative Learning Control (ILC) was proposed to control the accelerations. The results reveal that the tracking performance of MPIC can substantially be improved by ILC. In particular, the free vibration response is captured very accurately. The shape of the impact force is reasonable, and, as expected, exhibits a different shape for different parameters of the MKC system. Nevertheless, there are still several remaining challenges:

- A significant challenge in RTHS shock testing is effective control of the actuator to represent the response of the numerical substructure during the impact period of the shock event. The actuator, when struck by the pendulum, tends to move at a higher acceleration than required. Due to the delay of the amplifier ($\approx$1.5 ms), the compensator scheme is not able to counter this shaker recoil. To reduce this effect, an additional mass could be added on the shaker for higher inertia. Another idea is to use an additional sensor to determine the time instant of the hammer strike.
- To validate the RTHS results, a reference solution is of great value. This either involves an accurate modeling of the impact scenario (identification of contact parameters) or a full experimental test of this dynamical system.
- In the conducted RTHS test, only the acceleration is controlled. Any errors in the acceleration signal cumulate in the position error that will substantially influence the impact force. Hence, a position control loop should be added, which ideally uses the signals of an additional position sensor.

Future work will focus on resolving these issues.

References

1. American Association of State Highway and Transportation Officials: AASHTO LRFD Bridge Design Specifications. American Association of State Highway and Transportation Officials, Washington DC (2012)
2. Bartl, A.: Real-time hybrid substructure testing. PhD. Thesis, Technical University of Munich (2019)
3. Botelho, R.M.: Real-time hybrid substructuring for marine applications of vibration control and structural acoustics. Phd Thesis, University of Connecticut (2015)
4. Bristow, D.A., Tharayil, M., Alleyne, A.G.: A survey of iterative learning control: a learning-based method for high-performance tracking control. IEEE Control Syst. Mag. **3**, 96–114 (2006)
5. Carrion, J., Spencer, B.F.: Real-time hybrid testing using model-based delay compensation. In: 4th International Conference on Earthquake Engineering., Taipei, Tiawan, Paper No. 299, p. 10 (2006)
6. Carrion, J.E., Spencer Jr., B.F.: Model-based strategies for real-time hybrid testing. In: Newmark Structural Engineering Laboratory. Report Series No. 6. University of Illinois at Urbana-Champaign, Urbana, IL (2007)
7. Darby, A.P., Blakeborough, A., Williams, M.S.: Real-time substructure tests using hydraulic actuator. J. Eng. Mech. **125**(10), 1133–1139 (1999)
8. Department of the Navy (NAVY): MIL-S-901D: requirements for shock tests. H.I. (high-impact) shipboard machinery, equipment, and systems (1989)
9. Federal Emergency Management Agency (FEMA): Risk management series: reference manual to mitigate potential terrorist attacks against buildings (FEMA 426) (2003)
10. Fermandois, G.A., Spençer, B.F.: Model-based framework for multi-axial real-time hybrid simulation testing. Earthq. Eng. Eng. Vib. **16**, 671–691 (2017)
11. Harris, C.M., Piersol, A.G.: Harris' Shock and Vibration Handbook, 5th edn. McGraw-Hill, New York (2002)
12. Hochrainer, M.J., Puhwein, A.M.: Investigation of nonlinear dynamic phenomena applying real-time hybrid simulation. In Nonlinear Structures and Systems, vol. 1, pp. 125–131. Springer, Berlin (2020)
13. Horiuchi, T., Nakagawa, M., Sugano, M., Konno, T.: Development of a real-time hybrid experimental system with actuator delay compensation. In 11th World Conference on Earthquake Engineering, Paper No. 660 (1996)
14. Horiuchi, T., Inoue, M., Konno, T., Namita, Y.: Real-time hybrid experimental system with actuator delay compensation and its application to a piping system with energy absorber. Earthq. Eng. Struct. Dyn. **28**(10), 1121–1141 (1999)

15. Insam, C., Kist, A., Rixen, D.J.: High fidelity real-time hybrid substructure testing using iterative learning control. In ISR - 52nd International Symposium on Robotics. VDE Verlag, Berlin, Offenbach (2020)
16. Insam, C., Kist, A., Schwalm, H., Rixen, D.J.: Robust and high fidelity real-time hybrid substructuring. Mech. Syst. Signal Process. **157**, 107720 (2021)
17. Lang, G.F.: Electrodynamic shaker fundamentals. Sound Vibration **31**, 14–23 (1997)
18. Lin, F., Maghareh, A., Dyke, S.J., Lu, X.: Experimental implementation of predictive indicators for configuring a real-time hybrid simulation. Eng. Struct. **101**, 427–438 (2015)
19. Ljung, L.: System Identification: Theory for the User. Prentice Hall Information and System Sciences Series. Prentice Hall, Hoboken (1999)
20. Maghareh, A., Dyke, S.J., Silva, C.E.: A self-tuning robust control system for nonlinear real-time hybrid simulation. Earthq. Eng. Struct. Dyn. **49**(7), 695–715 (2020)
21. Nakashima, M., Masaoka, N.: Real-time on-line test for MDOF systems. Earthq. Eng. Struct. Dyn. **28**, 393–420 (1999)
22. Nakashima, M., Kato, H., Takaoka, E.: Development of real-time pseudo dynamic testing. Earthq. Eng. Struct. Dyn. **21**(1), 79–92 (1992)
23. Norrlöf, M.: Iterative learning control - analysis, design, and experiments. Ph.D. Thesis, Linköping University (2000)
24. Owens, D.H.: Iterative Learning Control, pp. 1–8. Springer, London (2014)
25. Owens, D., Daley, S.: Iterative learning control - monotonicity and optimization. Appl. Math. Comput. Sci. **18**, 279–293 (2008)
26. Phillips, B.M., Spencer, B.F.: Model-based feedforward-feedback actuator control for real-time hybrid simulation. J. Struct. Eng. **139**(7), 1205–1214 (2013)
27. Phillips, B.M., Spencer, B.F.: Model-based multiactuator control for real-time hybrid simulation. J. Eng. Mech. **139**(2), 219–228 (2013)
28. Ryschkewitsch, M.G.: Pyroshock test criteria. NASA-STD-7003A (2011)
29. Saouma, V., Sivaselvan, M.: Hybrid Simulation: Theory, Implementation and Applications. Taylor & Francis, Milton Park (2008)

Chapter 5
Band Gap Estimation of D-LEGO Meta-structures Using FRF-Based Substructuring and Bloch Wave Theory

Hrishikesh S. Gosavi, Phanisri P. Pratapa, and Vijaya V. N. Sriram Malladi

Abstract Periodic structures are found to exhibit band gaps which are frequency bandwidths where structural vibrations are absorbed. In this paper, meta-structures are built by dynamically linking oscillators in a periodic pattern, which are referred to as dynamically linked element grade oscillators or D-LEGOs. The location of the band gaps is numerically determined for a one-dimensional D-LEGO. The unit cell for the D-LEGO structure is considered to be made up of two longitudinal bar elements of different properties. For such a structure, the frequency response functions (FRFs) of a single unit cell are used to estimate the band gaps of a periodic-lattice structure by adapting the Bloch wave theory. Alternatively, the FRF of the multi-unit cell is determined using FRF-based substructuring (FBS) approach. The band gaps resulting from these two approaches are compared and verified.

Keywords Band gaps · Substructuring · Wave propagation · Periodic structure · Bloch wave theory

5.1 Introduction

Periodic structures have been a subject of extensive research in previous decades. This is due to the unique property demonstrated by periodic structures of inhibiting the propagation of waves (elastic waves in case of vibrations) through a structure in a frequency band, which is called as band gap. Band gaps were first observed in phononic crystals [1, 2]. Since then, the existence of band gaps has been proved numerous times through theoretical and experimental approaches. These band gaps are produced as a result of difference between the impedances which produces a Bragg scattering effect. By changing the spatial periodicity and the impedance mismatch, the frequency range of the band gap can be changed as per requirement [3]. The impedance mismatch can be achieved when the elements of a unit cell, which repeats throughout the periodic structure, have significantly different properties (material and/or geometrical properties). Changes in the properties of a unit cell change the range and location of band gaps.

Owing to the property of band gaps, periodic structures have wide applications as vibration isolators [4]. Periodic structures also have wide applications in aeronautics, space structures, optical filters, and shielding structures from earthquakes. Accurate estimation of elastic band gaps in a periodic structure hence becomes imperative for its proper implementation in any particular application. A number of methods have been proposed to compute the elastic band gaps in a periodic structure. These include the plane wave expansion (PWE) method, finite-difference time-domain (FDTD) method, lumped mass method, transfer-matrix method, and method based on the Bloch wave theory [5]. Due to its simplicity and good results, this paper utilizes the Bloch wave theory to estimate band gaps.

This paper discusses the estimation of elastic band gaps in a periodic structure consisting of D-LEGOs. The unit cell of this periodic structure is considered to be made up of two bars having different dimensions and material properties. Such unit cells are put together to form a periodic structure as shown in Fig. 5.1.

The finite element method is used to calculate the mass and stiffness matrices using one-dimensional bar elements. The mass and stiffness matrices are then used to estimate the band gaps in the structure using the Bloch wave theory. The FRF

H. S. Gosavi (✉) · V. V. N. S. Malladi
Michigan Technological University, Houghton, MI, USA
e-mail: hsgosavi@mtu.edu; smalladi@mtu.edu

P. P. Pratapa
Indian Institute of Technology, Madras, Chennai, India
e-mail: ppratapa@civil.iitm.ac.in

M. Allen et al. (eds.), *Dynamic Substructures, Volume 4*, Conference Proceedings of the Society for Experimental Mechanics Series,
https://doi.org/10.1007/978-3-031-04094-8_5

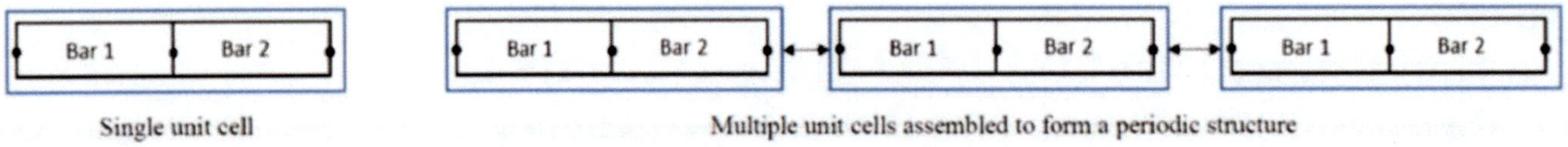

Fig. 5.1 Multiple unit cells, each consisting of two bars, were assembled to form a periodic structure

for the periodic structure is calculated by assembling multiple unit cells together using FRF-based substructuring [6–8]. The FRF of this multi-unit cell structure is then found to show a band gap. The range of this band gap is compared with that of the one found using the Bloch wave theory. For this study, the first bar is considered to be made of aluminum and the second one of polylactic acid (PLA). Both the bars are considered to be of half inch square cross section and of one inch length. These unit cells are assembled along their lengths to get a long periodic structure which results in longitudinal band gaps.

5.2 Bloch Wave Theory

The Bloch wave theory, also known as the Floquet-Bloch theory, analyzes the propagation of waves through a periodic structure [9, 10]. It relates the displacement of a particular unit cell with its previous unit cell through a dispersion constant. Consider a periodic structure of N unit cells as shown in Fig. 5.2.

Let us consider three consecutive unit cells, i.e., $(N-1)^{\text{th}}$, N^{th}, and $(N+1)^{\text{th}}$ unit cell. According to the Floquet-Bloch theory, the relation between the displacements of the nodes of two consecutive unit cells is given by:

$$\begin{bmatrix} U_3 \\ U_4 \end{bmatrix} = e^{i\mu} \begin{bmatrix} U_1 \\ U_2 \end{bmatrix}^{N-1} \text{ and } \begin{bmatrix} U_5 \\ U_6 \end{bmatrix}^{N+1} = e^{i\mu} \begin{bmatrix} U_3 \\ U_4 \end{bmatrix}^{N} \tag{5.1}$$

where $U_1 \ldots U_6$ are displacements of nodes from 1 to 6 and μ is the dispersion constant. Let $\lambda = e^{i\mu}$.

Consider the differential equation of the system:

$$M \left[\ddot{U}\right] + K \left[U\right] = 0 \tag{5.2}$$

where M and K are the mass and stiffness matrices for "N" unit cells. For a single one-dimensional bar element, the mass and stiffness matrices are given as:

$$M = \frac{\rho Al}{6} \begin{bmatrix} 2 & 1 \\ 1 & 2 \end{bmatrix} \& K = \frac{EA}{l} \begin{bmatrix} 1 & -1 \\ -1 & 1 \end{bmatrix} \tag{5.3}$$

For N unit cells, the mass and stiffness matrices from (5.3) are assembled. To calculate the dispersion characteristics, a single block from the mass and stiffness matrices is considered. From this single block, the stiffness matrix for $(N-1)^{\text{th}}$, N^{th}, and $(N+1)^{\text{th}}$ unit cell can be written as:

$$K_{N-1} = \frac{EA}{l} \begin{bmatrix} 0 & -1 \\ 0 & 0 \end{bmatrix} \qquad K_N = \frac{EA}{l} \begin{bmatrix} 2 & -1 \\ -1 & 2 \end{bmatrix} \qquad K_{N+1} = \frac{EA}{l} \begin{bmatrix} 0 & 0 \\ -1 & 0 \end{bmatrix} \tag{5.4}$$

and the mass matrix for these unit cells is:

$$M = \frac{\rho Al}{6} \begin{bmatrix} 2 & 1 \\ 1 & 2 \end{bmatrix} \tag{5.5}$$

Using Eqs. (5.1), (5.3), (5.4), (5.5), and (5.6), the combined stiffness matrix for $(N-1)^{\text{th}}$, N^{th}, and $(N+1)^{\text{th}}$ unit cells can be written as:

$$K = \frac{1}{\lambda} K_{N-1} + K_N + \lambda K_{N+1} \tag{5.6}$$

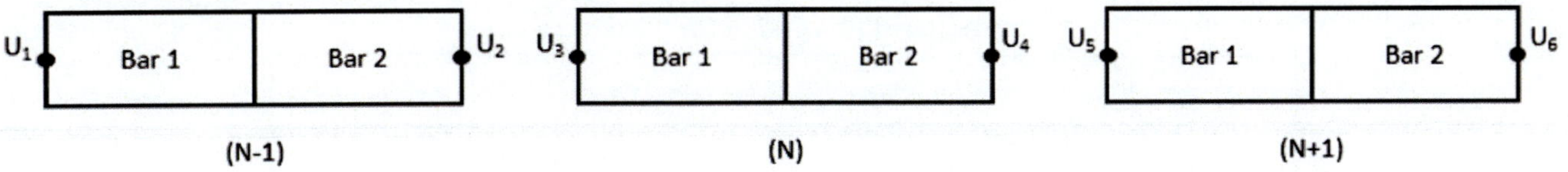

Fig. 5.2 Consider three consecutive unit cells $N-1$, N, and $N+1$

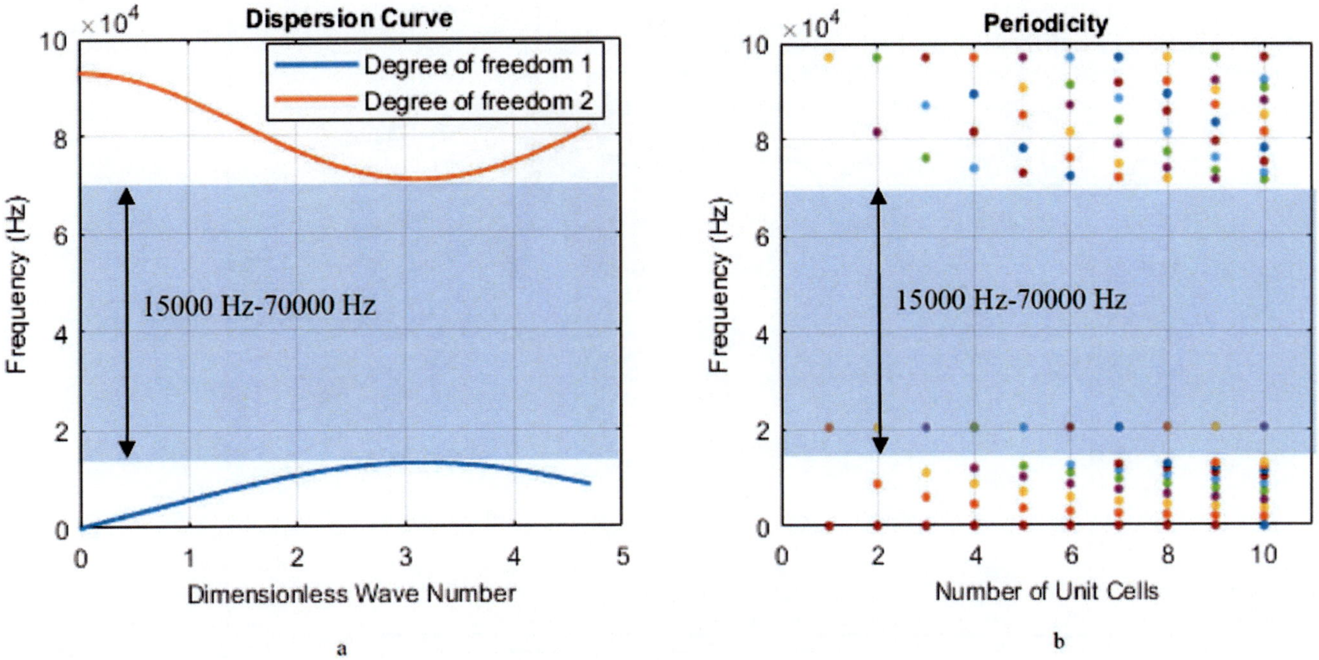

Fig. 5.3 The dispersion curve (**a**) and the periodicity (**b**) show a band gap from 15,000 to 70,000 Hz

Hence, Eq. (5.2) becomes:

$$M\left[\ddot{U}\right]+\left(\frac{1}{\lambda}K_{N-1}+K_N+\lambda K_{N+1}\right)[U]=0 \tag{5.7}$$

The corresponding eigenvalue problem can be formulated from (5.7). Solving the eigenvalue problem yields the frequency values which are plotted against the dispersion constant to get the dispersion curves. The dispersion curve is shown in Fig. 5.3.

In Fig. 5.3, the range 15,000–70,000 Hz does not have any frequency values which indicates the presence of a band gap. This frequency range should not have any eigenvalues or natural frequencies in the FRF of the multi-unit cell structure. The FRF of multi-unit cell structure is calculated using FRF-based substructuring in the next section.

5.3 FRF-Based Substructuring (FBS)

The FRF coupling analysis method was used to calculate the FRF for multi-unit cell periodic structure. This method combines the FRF data from subsystem models. Consider two subsystems "A" and "B" with internal degrees of freedoms "d" and shared degrees of freedom "s." The response matrix for both the systems is given by:

$$H_A=\begin{bmatrix} H_{dd}^A & H_{sd}^A \\ H_{ds}^A & H_{ss}^A \end{bmatrix} \& H_B=\begin{bmatrix} H_{dd}^B & H_{dd}^B \\ H_{dd}^B & H_{dd}^B \end{bmatrix} \tag{5.9}$$

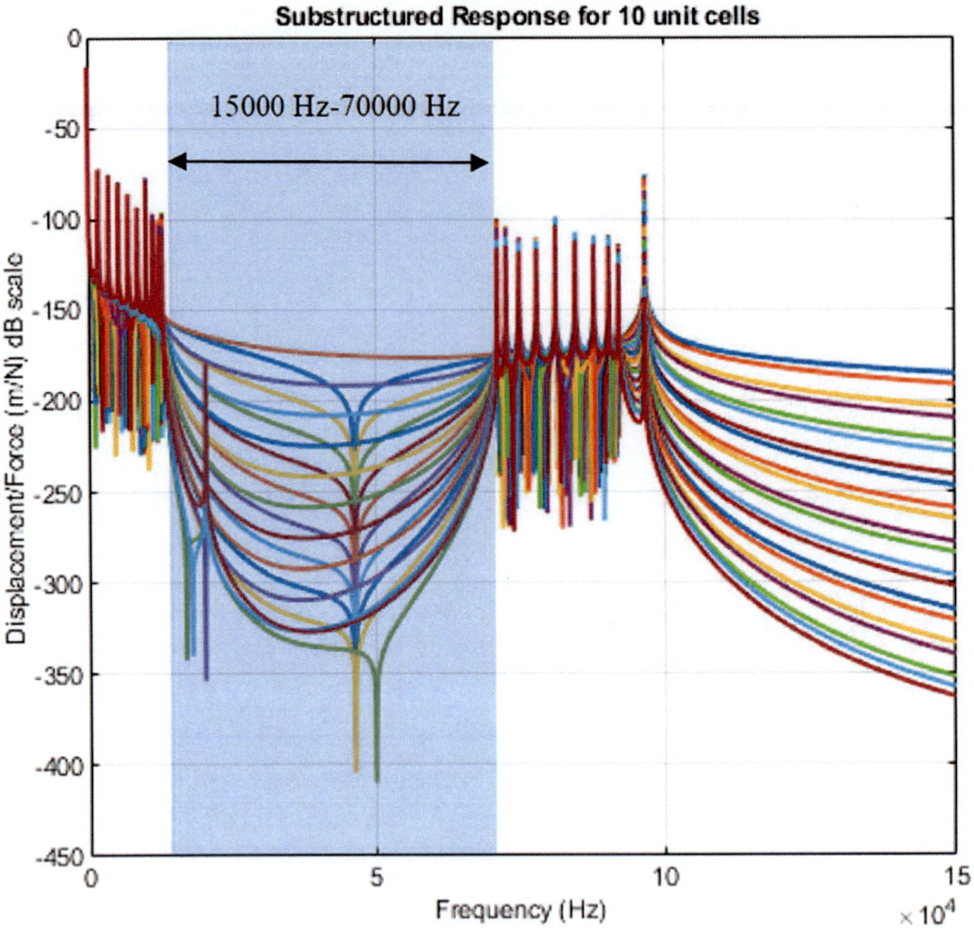

Fig. 5.4 The response for a periodic structure having ten unit cells was calculated using FBS which has the same band gap as Fig. 5.3

The response for the coupled system C is given by:

$$H_C = \left[[H_A]^{-1} \oplus \ [H_B]^{-1} \ \right]^{-1} \tag{5.10}$$

where ⊕ symbol indicates the assembly of matrices such that the shared degrees of freedom are added.

The response for multiple unit cells was found using coupling analysis method of FRF-based substructuring by considering each unit cell as a subsystem. The substructured response for ten unit cells is as shown in Fig. 5.4.

The FRF in the figure shows that there are no peaks in the range 15,000–70,000 Hz, which is a characteristic of a meta-structure with band gap. This frequency range is exactly the same as the one estimated using the Bloch wave theory. Hence, the band gap estimated using two approaches has been compared and verified.

5.4 Results for Different Number of Elements in the Finite Element Model

The results shown in Figs. 5.3 and 5.4 were calculated for one element in each bar. However, to confirm the accuracy of the results, we would need to calculate the responses and band gaps for higher number of elements in a bar. Figures below show the frequency response and the dispersion curves for 10, 15, and 20 number of elements in bar and for 10 number of unit cells.

For ten elements in both the bars of unit cell (Figs. 5.5 and 5.6):

For 20 elements in both the bars of the unit cell (Figs. 5.7 and 5.8):

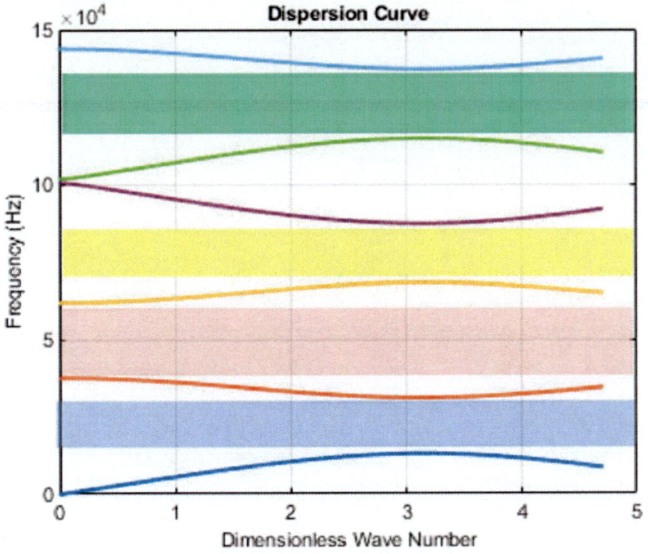

Fig. 5.5 Dispersion curve for ten elements

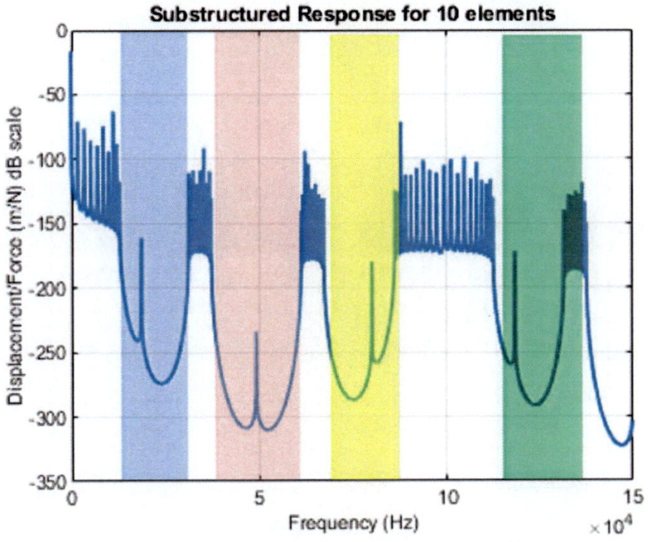

Fig. 5.6 Substructured response for ten elements

From the above figures, we observe that the band gaps are correctly verified by both the approaches for 10 and 20 number of unit cells. Hence, the accuracy of the results has been confirmed.

5.5 Conclusion

Thus, a periodic structure was formed by assembling multiple unit cells together which consisted of two bars made of different materials and cross section. Finite element model having one-dimensional bar elements was used to get the mass and stiffness matrices. These matrices were used in the Bloch wave method to estimate the band gaps in the periodic structure. The FRF of the periodic structure for multiple unit cells was then calculated using the FRF-based substructuring. This FRF was found to show band gap which had the same frequency range as the band gap estimated using the Bloch wave method. Hence, the band gaps from these two approaches were compared and found to match.

The location of band gap depends on the material and geometrical properties of the bars in the unit cell. These properties can thus be changed to have the band gaps in any intended frequency range for a particular application. The future scope of this work includes the estimation of band gaps for unit cells of beam elements which have longitudinal and rotational degrees of freedom and the experimental verification of the band gaps estimated in this study.

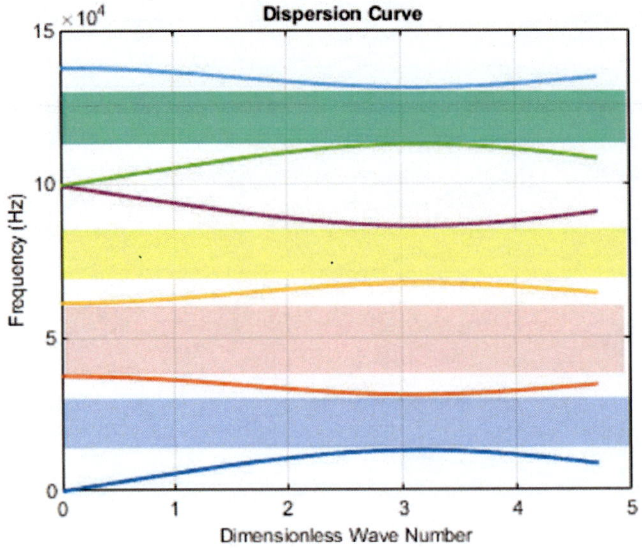

Fig. 5.7 Dispersion curve for 20 elements

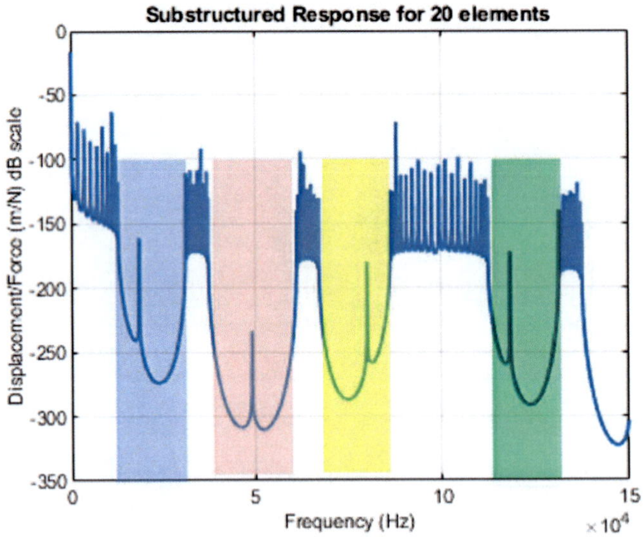

Fig. 5.8 Substructured response for 20 elements

References

1. Yan, Z.-Z., Wang, Y.-S.: Wavelet-based method for calculating elastic band gaps of two-dimensional phononic crystals. Phys. Rev. B. **74**, 328 (2006)
2. Hussein, M.I., Leamy, M.J., Ruzzene, M.: Dynamics of phononic materials and structures: historical origins, recent progress, and future outlook. Appl. Mech. Rev. **66** (2014)
3. Nobrega, E.D., Gautier, F., Pelat, A., DosSantos, J.M.C.: Vibration band gaps for elastic metamaterial rods using wave finite element method. Mech. Syst. Signal Process. **79**, 192–202 (2016)
4. Li-Jie, W., Song, H.-W.: Band gap analysis of periodic structures based on cell experimental frequency response functions (FRFs). Acta Mech. Sinica. **35**, 156–173 (2019)
5. Cheng, Z.-B., Shi, Z.-F.: Multi-mass-spring model and energy transmission of one-dimensional periodic structures. IEEE Trans. Ultrasonics Ferroelectr. Freq. Contr. **61**(5), 739–746 (2014)
6. Drozg, A., Cepon, G., Boltezar, M.: Full-degrees-of-freedom frequency based substructuring. Mech. Syst. Signal Process. **98**, 570–579 (2017)
7. de Klerk, D., Rixen, D.J., de Jong, J.: The frequency based substructuring (FBS) Method reformulated according to dual domain decomposition method. In: Proceedings of the XXIV international modal analysis conference, St. Louis, MO (2006)
8. Wyckaert, K., Xu, K.Q., Mas, P.: The virtues of static and dynamic compensation for FRF based substructuring. Proc. SPIE Int. Soc. Optical Eng. **3089**, 1463–1468 (2000)

9. Cheung, S.-K., Chan, T.-L., Zhang, Z.-Q., Chan, C.T.: Large photonic band gaps in certain periodic and quasiperiodic networks in two and three dimensions'. Phys. Rev. B. **70**, 125104 (2004)
10. Hoang, V., Plum, M., Wieners, C.: A computer-assisted proof for photonic band gaps. J. Appl. Math. Phys. **60**, 1035–1062 (2009)

Chapter 6
An Assessment on the Efficiency of Different Reduction Techniques Based on Substructuring for Bladed Disk Systems with Shrouds

Ehsan Naghizadeh and Ender Cigeroglu

Abstract Vibration analysis of shrouded bladed disk systems often becomes expensive due to friction nonlinearities and randomness stemming from mistuning phenomena. This implies a great demand for a highly efficient model order reduction technique to not only reduce the computational effort but, more importantly, provide reliable displacement predictions on certain degrees of freedom (shrouds). The latter becomes more critical in bladed disks with shroud contacts, since the promising results from contact models are limited by the accuracy of displacements predicted by reduced-order models for shroud degrees of freedom. In this study, some notable reduction of order methodologies based on substructuring, namely, fixed interface (Craig-Bampton), free interface (Rubin), and dual Craig-Bampton, and the mixed interface method, which is a combination of free and fixed interface methods, are investigated. The center of attention in this work is the modal contributions of components to the final result and the influence of modal characteristics of substructures on the efficiency of a particular reduction technique. To this end, the methods are examined by a different number of retained modes. The effect of adding up more vibration modes to the reduction basis on the accuracy and computational cost for each reduction technique is compared for predefined error tolerance. It is concluded that the physical characteristics of the blade and disk components significantly affect the forced response of the bladed disk system. Consequently, it can be capitalized on to find a more effective reduction technique for the specific geometry of shrouded blisks to address high computational cost and accurate forced response required in specific areas.

Keywords Reduction of order techniques · Substructuring · Bladed disk systems · Free and fixed interface methods

6.1 Introduction

Bladed disks are delicate mechanical parts that require in-depth investigation to predict their dynamic behavior due to their unique geometry. Scientists have tried different modeling strategies from lumped parameters [1] to elaborate finite element models. While the former was more convenient from a computational point of view, the latter requires much more time and storage, especially when the nonlinearity is considered. The primary source of nonlinearity in bladed disk systems is the friction between shrouds, blades and disk, or blades and underplatform dampers. Besides the nonlinearity, which is the reason for the need for a much elaborate finite element model [2], mistuning, due to its random nature, obligates conducting a statistical study that involves solving the problem of a considerable number of times [2]. The immense computational cost dictated by the situation mentioned above is overwhelming even for today's computers [3]. Another critical issue to be considered is the limiting act of reduction techniques on the final results of the nonlinear analysis. As asserted by Petrov [4], the contact model needs to be feed in with accurate information about the displacements of the contact surfaces; otherwise, regardless of the robustness of the contact model, it will be unable to predict the system behavior accurately. As a result, the search for an efficient reduction technique to be able to make the balance between computational cost and required accuracy is an ongoing topic of research.

Reduction of order, based on substructuring, is one of the well-known methods in structural dynamics [5]. This approach usually consists of three main steps, dismantling the structure to several simpler substructures, reducing the size of each substructure, and interconnecting the substructures to reach the reference structure. After the reduction phase, each substructure is described with some generalized DOFs as well as master DOFs. Master DOFs usually correspond to DOFs

---

E. Naghizadeh · E. Cigeroglu (✉)
Department of Mechanical Engineering, Graduate School of Natural and Applied Sciences, Middle East Technical University, Ankara, Turkey
e-mail: naghizadeh.ehsan@metu.edu.tr; ender@metu.edu.tr

M. Allen et al. (eds.), *Dynamic Substructures, Volume 4*, Conference Proceedings of the Society for Experimental Mechanics Series,
https://doi.org/10.1007/978-3-031-04094-8_6

of which physical information is required, i.e., force or displacement. Generalized DOFs, however, as the name implies, are related to the modal domain.

Reduction of order techniques can be classified concerning various measures. One of the main criteria for distinguishing the reduction techniques is how they describe each substructure's interface [6]; according to this, "fixed interface methods" and "free interface methods" were introduced to the literature [7–9]. Both methods use a combination of static and dynamic modes to build a reduction basis for each substructure. For the fixed interface methods, vibration modes are derived, while the boundary DOFs are held fixed, and static constraint modes are employed to describe the interaction of the substructure with others. On the other hand, interface DOFs are left free in calculating the dynamic modes, and residual flexibility modes carry the burden of representing the interface in the free interface methods. While both approaches have their merits and demerits, the question of "which method suits the best for a specific problem?" is of great concern.

A considerable amount of research has been conducted to compare the two strategies for the vibration of bladed disk systems in recent years. Mashayekhi et al. [10, 11] compared the most famous reduced order method of each category. Their studies only consider the blade sector, which is not irrelevant since the system's vibration mainly occurs at the blade. Pourkiaee and Zucca [12] proposed a method for the reduction of bladed disk systems with shrouds. Unlike [10, 11], they consider blade and disk components, but the prime novelty is to use loaded interface (LI) mode shapes, originally introduced in [13], to further reduce the system by eliminating the blade-to-disk interface DOFs. Moreover, Quaegebeur et al. [14, 15] propose a two-stage reduction, in which a new reduction basis is devised based on the cyclic property of the system. Yuan et al. [16] recently come up with an adaptive reduction method for systems with non-conservative forces. They implemented the technique to solve a forced vibration problem of bladed disk systems with a dovetail joint.

In this paper, a comparative study is conducted to assess the efficiency of different reduction techniques for the specific geometry of the bladed disk systems with shrouds. From the fixed interface method, Hurty/Craig-Bampton [7]; free interface, Rubin [9] and dual Craig-Bampton [17] methods; and also a more recent method called "mixed interface method" [18], which is a proper generalization of the Craig-Bampton and dual Craig-Bampton method, are examined. The prime purpose of this work is to investigate the relation between the physics of each substructure (blade and disk) and their modal contributions to the final result after coupling. To this end, errors of the methods in predicting natural frequencies are compared with a new perspective. That is, how fast the error converges to zero when more modes are added to the reduction basis of each substructure when different types of modes are used in the reduction basis. Moreover, the computational time is as well compared between these methods.

6.2 Theory

In this section, a rather basic description of the reduction techniques, included in this study, is given. In order to implement the methods, a relatively elaborate finite element model, as illustrated in Fig. 6.1, is created on a commercial finite element software. It should also be noted that the system is considered to be tuned. Thus, besides all the reduction methods, the cyclic symmetry property is used to reduce the computation of the entire system to just a single sector. For the sake of simplicity, the first harmonic of the cyclic system is examined, in which all the blades vibrate in phase and with the same amplitudes.

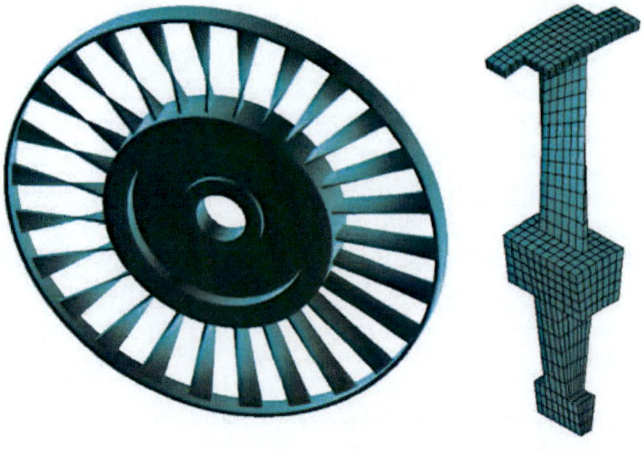

Fig. 6.1 FEM model of the bladed disk and a single sector

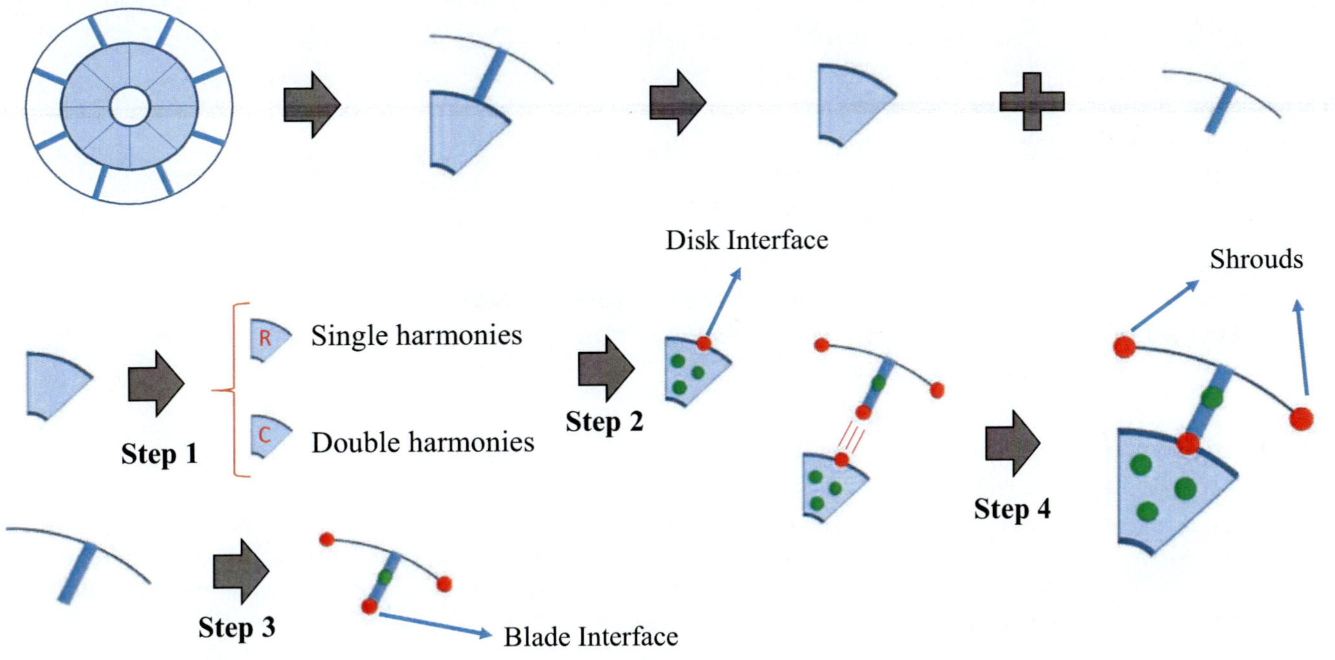

Fig. 6.2 Flowchart of the reduction procedure

Moreover, the disk-to-disk connection DOFs are not necessary to keep because the problem is going to be solved on the harmonic level. In Fig. 6.2, a flowchart of the reduction approach is provided. Although completely different in the reduction basis and performances, all the methods follow the same procedure described in Fig. 6.2.

6.2.1 Fixed Interface Methods

Although the early Guyan method [19] can be considered as a fixed interface method, this paper considers the famous Craig-Bampton method as the only representative of the fixed interface reduction methods.

In the Craig-Bampton method, two sets of mode shapes, i.e., static constraint modes (Λ_c) and fixed interface modes (ϕ_b), are used to create a reduction basis.

Static constraint modes are derived by neglecting the acceleration or dynamic motion of the system. Thus, the modes are analogous to static deformation of the system, while unit displacement is applied to one of the boundary DOFs, whereas other boundary DOFs are held fixed. The first step to calculate the static constraint modes is dividing the component DOFs into internal (i) and interface or boundary (b)DOFs.

$$\mathbf{u} = \begin{Bmatrix} \mathbf{x}_i \\ \mathbf{x}_b \end{Bmatrix}, \tag{6.1}$$

$$\mathbf{x}_i^{static} = -\mathbf{K}_{ii}^{-1}\mathbf{K}_{ib}\mathbf{x}_b \quad \rightarrow \quad -\mathbf{K}_{ii}^{-1}\mathbf{K}_{ib} = \boldsymbol{\Lambda}_c, \tag{6.2}$$

where $\mathbf{x}_i$ and $\mathbf{x}_b$ are displacement vectors of interior and boundary DOFs, respectively, and $\mathbf{K}$ is the stiffness matrix of the fundamental sector. The columns of Λ_c are the static constraint modes; then:

$$\begin{Bmatrix} \mathbf{x}_i \\ \mathbf{x}_b \end{Bmatrix} = \begin{bmatrix} \boldsymbol{\Lambda}_c \\ \mathbf{I} \end{bmatrix} \mathbf{x}_b. \tag{6.3}$$

Fixed interface modes are indeed vibration modes, while all the boundary DOFs are fixed. Consequently, they are easily driven from the eigenvalue problem as follows:

$$\left(\mathbf{K}_{ii} - \omega^2\mathbf{M}_{ii}\right)\boldsymbol{\phi}_b = \mathbf{0}. \tag{6.4}$$

Then, while the boundary displacements are kept as they are, the interior DOFs are approximated by some truncated generalized modal coordinates.

$$\mathbf{u} = \begin{Bmatrix} \mathbf{x}_i \\ \mathbf{x}_b \end{Bmatrix} \cong \begin{bmatrix} \boldsymbol{\phi}_b & \boldsymbol{\Lambda}_c \\ \mathbf{0} & \mathbf{I} \end{bmatrix} \begin{Bmatrix} \boldsymbol{\eta}_b \\ \mathbf{x}_b \end{Bmatrix}. \tag{6.5}$$

From the above expression, it is evident that the primary reduction actually takes place in reducing the interior DOFs by using modal truncation.

6.2.2 *Free Interface Method*

In this subsection, the dual Craig-Bampton method, together with the Rubin method, is investigated. While both methods use the same reduction basis, there are significant differences in representing the boundary and the procedure of coupling two substructures in these methods. In the following, a brief description of the reduction basis of both methods, together with the differences between fixed and free interface reduction basis, is provided.

The two main modes that are used in the free interface methods are attachment modes (Λ_a) or residual attachment modes (Λ_r) and free vibration modes (ϕ_r).

Attachment modes are a set of static modes which are used with free interface modes. Unlike constraint modes, there is no need to classify the component DOFs. However, forces at the boundary of the substructure are included in the overall displacement vector. The reason behind this is the need for this information (boundary forces) for the very act of coupling substructures.

$$\mathbf{u} = \begin{Bmatrix} \mathbf{x} \\ \mathbf{f}_b \end{Bmatrix}, \tag{6.6}$$

where $\mathbf{f}_b$ is the vector of forces acting on the boundary DOFs. Then the static equations of motion can be derived as:

$$\begin{bmatrix} \mathbf{K} & -\mathbf{A}^T \\ -\mathbf{A} & \mathbf{0} \end{bmatrix} \begin{Bmatrix} \mathbf{x} \\ \mathbf{f}_b \end{Bmatrix} = \begin{Bmatrix} \mathbf{f} \\ \mathbf{0} \end{Bmatrix} \tag{6.7}$$

From the above equation, $\mathbf{u}^{\text{static}}$ can be calculated from $\mathbf{f}_i$:

$$\mathbf{u}^{\text{static}} = \mathbf{K}^{-1}\mathbf{A}^T\mathbf{f}_b = \boldsymbol{\Lambda}_a\mathbf{f}_b. \tag{6.8}$$

$\mathbf{K}^{-1}$ is the flexibility matrix, post-multiplication of which by the Boolean matrix, $\mathbf{A}^T$, picks the columns associated with the boundary DOFs. Hence, the attachment modes (Λ_a) can be considered as static deformations due to a unit force applied to a specific boundary DOF.

While the attachment modes are best suited for experimental substructuring, since the experimental validation of them are more convenient, in computational substructuring, these modes can be further refined to reach a much more satisfactory reduction basis [20].

The idea here comes from noticing that the truncated vibration modes and attachment modes share similar information when the flexibility matrix expands onto the mode shapes of the system. As it can be understood from Eq. (6.9), including the same modes in the reduction basis causes them to be not orthogonal with **K** and **M** matrices. In order to have an orthogonal reduction basis, one may subtract the contribution of retained vibration modes from $\mathbf{K}^{-1}$ matrix, creating the residual flexibility matrix (Λ_r).

$$\boldsymbol{\Lambda}_r = \left(\mathbf{K}^{-1} - \sum_{i=1}^{n_r} \frac{\boldsymbol{\phi}_i\boldsymbol{\phi}_i^T}{\omega^2}\right)\mathbf{A}^T. \tag{6.9}$$

In the above equation, n_r is the number of retained vibration modes in the reduction basis.

Free interface modes are the vibration modes, while the structure is free from the interface, which can be defined by the following eigenvalue problem:

$$\left(\mathbf{K} - \omega^2\mathbf{M}\right)\boldsymbol{\phi}_r = \mathbf{0} \tag{6.10}$$

Rubin Method

In the Rubin method, the displacement vector of the substructure is expressed by the combination of a static solution by using the residual attachment modes and a dynamic solution from free interface vibration modes [9].

Note that in this paper we did not consider the distinction between the rigid body modes and vibration modes, providing that the eigensolver of interest is able to derive the orthogonal rigid body modes from the eigenvalue problem. Otherwise, the calculation of the rigid body modes should be taken care of separately (see [19]). Then, displacement vector can be written as follows:

$$\mathbf{x} \cong [\boldsymbol{\Phi}_r \ \ \boldsymbol{\Lambda}_r]\begin{Bmatrix}\boldsymbol{\eta}_r \\ \mathbf{f}_b\end{Bmatrix}. \tag{6.11}$$

The combination of the modal coordinates and the boundary forces leads to the so-called dual representation. In the Rubin method however, in order to facilitate the coupling of the substructures, the forces at the boundary again are transformed to the boundary displacements in a procedure described as follows:

$$\mathbf{x}_b = \mathbf{A}\mathbf{x} = \mathbf{A}\left(\boldsymbol{\Lambda}_r\mathbf{f}_b + \boldsymbol{\Phi}_r\boldsymbol{\eta}_\mathbf{r}\right) \tag{6.12}$$

Replacing the boundary forces in Eq. (6.11) by corresponding displacements derived from Eq. (6.12), one can obtain:

$$\begin{Bmatrix}\boldsymbol{\eta}_r \\ \mathbf{f}_b\end{Bmatrix} = \begin{bmatrix}\mathbf{I} & \mathbf{0} \\ (\mathbf{A}\boldsymbol{\Lambda}_r)^{-1}\mathbf{A}\boldsymbol{\phi}_r & (\mathbf{A}\boldsymbol{\Lambda}_r)^{-1}\end{bmatrix}\begin{Bmatrix}\boldsymbol{\eta}_r \\ \mathbf{x}_b\end{Bmatrix}. \tag{6.13}$$

As is evident from Eq. (6.13), the final coordinates are some modal coordinates beside the displacements at the boundary.

Dual Craig-Bampton Method

The same reduction basis is used for the dual Craig-Bampton method; however, instead of performing another transformation to reach to the interface displacements, forces are used to couple the substructures.

First, the boundary forces are included in the substructure's coordinates, and the equation of motion is rewritten as:

$$\begin{bmatrix}\mathbf{M} & \mathbf{0} \\ \mathbf{0} & \mathbf{0}\end{bmatrix}\begin{Bmatrix}\ddot{\mathbf{x}} \\ \ddot{\mathbf{f}}_b\end{Bmatrix} + \begin{bmatrix}\mathbf{K} & -\mathbf{L} \\ -\mathbf{L} & \mathbf{0}\end{bmatrix}\begin{Bmatrix}\mathbf{x} \\ \mathbf{f}_b\end{Bmatrix} = \begin{Bmatrix}\mathbf{p} \\ -\mathbf{x}_b\end{Bmatrix}. \tag{6.14}$$

Then one can write the reduction transformation as:

$$\begin{Bmatrix}\mathbf{x} \\ \mathbf{f}_b\end{Bmatrix} = \begin{bmatrix}\boldsymbol{\phi}_r & \boldsymbol{\Lambda}_r \\ \mathbf{0} & \mathbf{I}\end{bmatrix}\begin{Bmatrix}\boldsymbol{\eta} \\ \mathbf{f}_b\end{Bmatrix}. \tag{6.15}$$

6.2.3 *Mixed Interface Method*

The mixed interface method can be considered as a general case in which both the Craig-Bampton and dual Craig-Bampton methods can coexist. In this method, a combination of the constraint and attachment modes is used to describe the boundary of the substructure. The aim is to define a portion of the interface DOFs with free interface modes, while other DOFs are represented by fixed interface modes.

The procedure starts with partitioning the boundary coordinates of the substructure to free and fixed. Consequently, vibration modes are calculated by fixing only a portion of the boundary DOFs.

As in the case of the dual Craig-Bampton method, forces corresponding to the DOFs which are going to be represented as free are added to the coordinates of the substructure.

$$\mathbf{q} = \begin{Bmatrix} \mathbf{x}_i \\ \mathbf{x}_{\text{free}} \\ \mathbf{x}_{\text{fixed}} \\ \mathbf{f}_{\text{free}} \end{Bmatrix}. \tag{6.16}$$

Then the displacement vectors $\mathbf{x}_i$ and $\mathbf{x}_{\text{free}}$ are approximated, while the others are kept in the final coordinates of the substructure. According to [18], the approximation is as follows:

$$\begin{Bmatrix} \mathbf{x}_i \\ \mathbf{x}_{\text{free}} \end{Bmatrix} = \mathbf{x}_m \cong \boldsymbol{\phi}_m \boldsymbol{\eta}_m + \boldsymbol{\Lambda}_r \mathbf{f}_{\text{free}} + \boldsymbol{\Lambda}_c \mathbf{x}_{\text{fixed}}, \tag{6.17}$$

while

$$\left(\mathbf{K}_{mm} - \omega^2 \mathbf{M}_{mm}\right) \boldsymbol{\phi}_m = \mathbf{0}. \tag{6.18}$$

The residual attachment, Λ_r, and constraint modes, Λ_c, can be computed similar to the procedure in the previous subsections.

6.3 Numerical Results

In this section, numerical results obtained from each of the reduction methods are compared by retaining different number of modes in order to observe how accurate each method can predict the natural frequencies and mode shapes of the cyclically symmetric bladed disk system. The information about the size of the model in terms of number of degrees of freedom (DOFs) is demonstrated in Table 6.1. In Table 6.2, a description of the different reduction techniques is given. In the case of disk section, there is only one interface area; on the blade however, there are blade-to-disk and shroud interfaces. The way in which the methods describe each interface is listed in Table 6.2. As is shown in the table, the CB method is the only method in which the blade-to-disk boundary is kept fixed on the disk component; later, it is going to be revealed that this is actually not the best choice. Moreover, the blade interfaces are described in three different ways, i.e., all free (DCB and Rubin), all fixed (CB), and mixed method. Another important information provided is the largest matrix to be inverted (L.M.I.) for each method. This is usually needed to derive the static modes. The issue becomes critical when the structure is free floating, in which the inversion of the stiffness matrix cannot be done conventionally. In our case study, in particular, this happens when the blade-to-disk interface and shroud interfaces are both considered to be free on the blade sector. To observe the effect of this phenomena on the computational burden of the problem, all the reductions are carried out on a personal computer, and the computational times are stored in Table 6.2. It's evident that, due to singularity problem in the calculation of the residual attachment modes for DCB and Rubin methods, they require significantly more time in comparison to the two other methods.

Since the computational burden of each method is perceived. Next, the methods are compared based on their accuracy in the determination of natural frequencies and mode shapes of the bladed disk system. The objective is to estimate the mode shapes and natural frequencies of the full system. Particularly, 12 lowest natural frequencies and corresponding mode shapes are obtained by each method. Results obtained are compared with the solutions of the full system (without reduction) solved on a commercial finite element software.

Table 6.1 Size of the FEM model

Model	Bladed disk system with shrouds (number of blades = 26)					
Segment	Blade	Disk	Sector	Full model	B-D interface	Shrouds
Size (DOFs)	2010	2478	4446	106,392	42	96

Table 6.2 Important characteristics of the methods

Method	Blade interface	Disk interface	Shrouds	L.M.I.[a]	Final DOFs	Computational time[b] (s)
CB	Fixed	Fixed	Fixed	2436	NB + ND B-D interface = 42 (disp.) Shrouds = 96 (disp.)	6.67
DCB	Free	Free	Free	2478[c]	NB[d] + ND B-D interface = 42 (force) Shrouds = 96 (force)	45.35
Rubin	Free	Free	Free	2478[c]	NB[d] + ND B-D interface = 42 (disp.) Shrouds = 96 (disp.)	47.56
MXD	Free	Free	Fixed	2436	NB + ND B-D interface = 42 (force) Shrouds = 96 (disp.)	6.57

NB and ND are the number of mode retained for blade and disk, respectively
[a]Largest matrix to be inverted (size)
[b]Time is calculated by running the code 3 times by considering 12 modes per subsystems and getting the average
[c]For the blade case since the stiffness matrix is singular, its invers is obtained by modal superposition
[d]Six rigid body modes are also added to the number of the modes considered for the blade

At first glance, it's evident from Fig. 6.3 that the way the methods identify the interfaces definitely affects the error in each method regardless of which method performs better. In terms of accuracy, the Rubin method clearly outperforms others, whereas the mixed method predicts the natural frequencies with higher error. Nevertheless, the pattern of change in error gives important information. First, the sudden decreases in the amount of the error in the sixth and the ninth natural frequencies suggest that the disk sector is actually contributing to the vibration of the whole system and the contribution is indispensable. Moreover, since the addition of the disk modes cannot improve the results in the CB method, it can be concluded that free vibration modes of the disk sector are a better choice for the reduction of the disk.

For the case of mode shapes, the same approach is used for the comparison. It should be noted that DCB method is exempted from this comparison, because it considers the forces at shroud boundary not the displacement. According to Fig. 6.4, for CB and Rubin methods, the mode shapes of the structure are predicted with a good accuracy. Small errors may stem from the error in the prediction of natural frequencies. On the other hand, the mixed method struggles to predict the mode shapes.

6.4 Conclusion

During this study, differences in the several reduction techniques, mainly in how they describe the substructure interfaces, are examined from a new point of view, that is, how each substructure contributes to the overall vibration and mode shapes of the entire system. To this end, an academic finite element model is created and used as the case study to examine the ability of the methods in terms of predicting natural frequencies and mode shapes of the model. Conclusion on the behavior of different methods and how adding more modes to their reduction basis affects their efficiency are as follows:

- The type of vibration modes (weather they are derived with fixed or free interfaces) severely affects the final accuracy of a reduction technique.
- Although the contribution of the disk vibration to the overall vibration of the system near its fundamental frequency is negligible, however, as soon as the sixth or higher natural frequencies are reached, the effect of adding the vibration modes of the disk sector to the reduction basis can be clearly seen.
- The free interface methods demonstrate very good accuracy; however, in the case of existing rigid body motion, the computational cost is significantly more.
- The mixed interface method struggles to compete with its rivals, but from another point of view, it could be a proper approach to circulate the need for dealing with singularity problem in the case of rigid body motion.

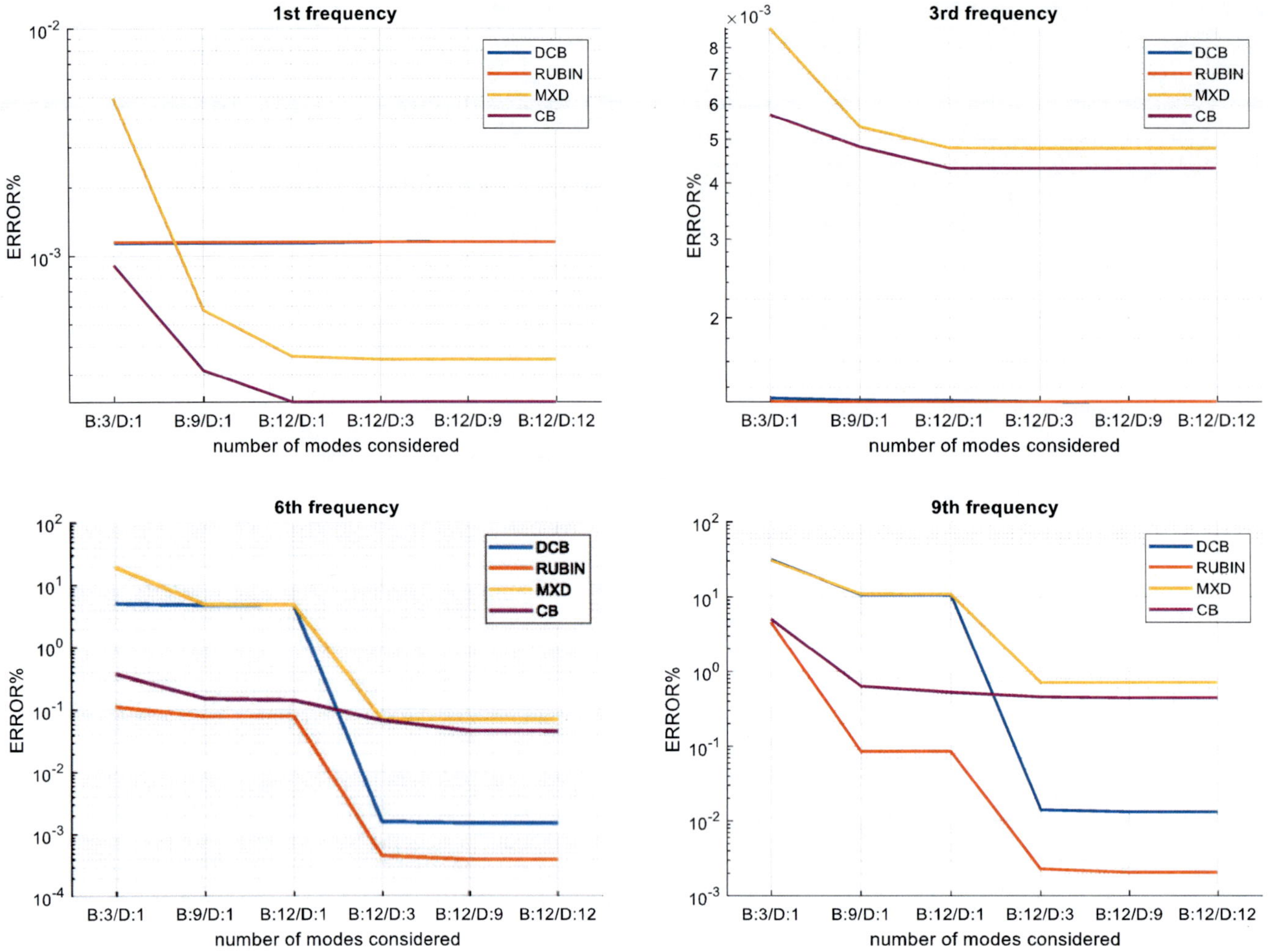

Fig. 6.3 Error criteria in predicting the natural frequencies by each method

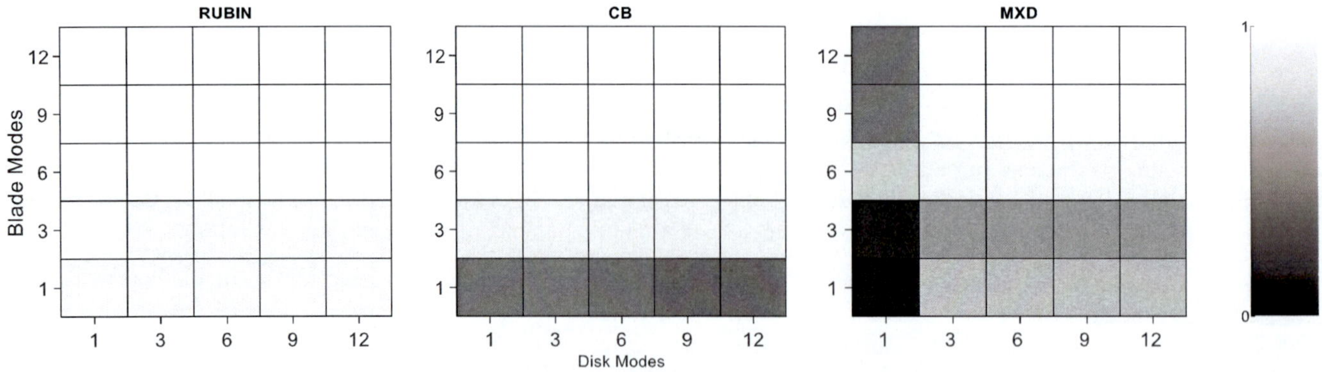

Fig. 6.4 Mac number associated with the ninth vibration mode of the system

References

1. Wagner, J.T.: Coupling of turbomachine blade vibrations through the rotor. J. Eng. Gas Turbines Power. **89**(4), 502–512 (1967). https://doi.org/10.1115/1.3616718
2. Bladh, C.M.P.: Component-mode-based reduced order modeling techniques for mistuned bladed disks-Part 1: theoretical models. J. Eng. Gas Turbines Power. **123**(1), 89–99 (2001). https://doi.org/10.1115/1.1338947

3. Rzadkowski, R.: The general model of free vibrations of mistuned bladed discs, part I: theory. J. Sound Vib. **173**(3), 377–393 (1994). https://doi.org/10.1006/JSVI.1994.1236
4. Petrov, E.P.: A high-accuracy model reduction for analysis of nonlinear vibrations in structures with contact interfaces. J. Eng. Gas Turbines Power. **133**(10), 1–10 (2011). https://doi.org/10.1115/1.4002810
5. Allen, M.S., Rixen, D., van der Seijs, M., Tiso, P., Abrahamsson, T., Mayes, R.L.: Substructuring in Engineering Dynamics, vol. 594. Springer, New York (2020)
6. De Klerk, D., Rixen, D.J., Voormeeren, S.N.: General framework for dynamic substructuring: history, review, and classification of techniques. AIAA J. **46**(5), 1169–1181 (2008). https://doi.org/10.2514/1.33274
7. Craig, R., Bampton, M.: Coupling of substructures for dynamic analyses to cite this version: HAL Id: hal-01537654 coupling of substructures for dynamic analyses. AIAA J. **6**(7), 1313–1319 (1968)
8. MacNeal, R.H.: A hybrid method of component mode synthesis. Comput. Struct. **1**(4), 581–601 (1971). https://doi.org/10.1016/0045-7949(71)90031-9
9. Rubin, S.: Improved component-mode representation for structural dynamic analysis. AIAA J. **13**(8), 995–1006 (1975). https://doi.org/10.2514/3.60497
10. Mashayekhi, F., Zucca, S., Nobari, A.S.: A comparison of two reduction techniques for forced response of shrouded blades with contact interfaces. **4**, 79–88 (2018). https://doi.org/10.1007/978-3-319-74654-8_7
11. Mashayekhi, F., Zucca, S., Nobari, A.S.: Evaluation of free interface-based reduction techniques for nonlinear forced response analysis of shrouded blades. Proc. Inst. Mech. Eng. Part C J. Mech. Eng. Sci. **233**(23–24), 7459–7475 (2019). https://doi.org/10.1177/0954406219872523
12. Mehrdad Pourkiaee, S., Zucca, S.: A reduced order model for nonlinear dynamics of mistuned bladed disks with shroud friction contacts. J. Eng. Gas Turbines Power. **141**(1) (2019). https://doi.org/10.1115/1.4041653
13. Benfield, W.A., Hruda, R.F.: Vibration analysis of structures by component mode substitution. AIAA J. **9**(7), 1255–1261 (1971). https://doi.org/10.2514/3.49936
14. Quaegebeur, S., Chouvion, B., Thouverez, F.: Nonlinear cyclic reduction for the analysis of mistuned cyclic systems. J. Sound Vib. **499** (2021). https://doi.org/10.1016/j.jsv.2021.116002
15. Quaegebeur, S., Chouvion, B., Thouverez, F., Berthe, L.: On a new nonlinear reduced-order model for capturing internal resonances in intentionally mistuned cyclic structures. J. Eng. Gas Turbines Power. **143**(2) (2021). https://doi.org/10.1115/1.4049138
16. Yuan, J., Schwingshackl, C., Wong, C., Salles, L.: On an improved adaptive reduced-order model for the computation of steady-state vibrations in large-scale non-conservative systems with friction joints. Nonlinear Dyn. **103**(4), 3283–3300 (2021). https://doi.org/10.1007/s11071-020-05890-2
17. Rixen, D.J.: A dual Craig-Bampton method for dynamic substructuring. J. Comput. Appl. Math. **168**(1–2), 383–391 (2004). https://doi.org/10.1016/j.cam.2003.12.014
18. Matichard, F., Gaudiller, L., Voormeeren, S.N., Van Der Valk, P.L.C, Rixen, D.J.: A general mixed boundary model reduction method for component mode synthesis Related content Hybrid modal nodal method for multibody smart structure model reduction: application to modal feedback control. https://doi.org/10.1088/1757-899X/10/1/012116
19. Géradin, M., Rixen, D.J.: Mechanical Vibrations: Theory and Application to Structural Dynamics. Wiley, Hoboken (2015)

Chapter 7
Control-Free Dynamic Substructuring of a Three-Storey Building

Sandor Beregi, David A. W. Barton, Djamel Rezgui, and Simon A. Neild

Abstract In this study, the dynamic substructuring of a benchtop three-storey building is realised where the two top storeys of the building are in real-time interaction with a simulated storey at the bottom of the structure. The challenging nature of this task is demonstrated by calculating the very limited stabilisable parameter domain if displacement-feedback control is used to connect the physical and virtual parts of the experiment. Instead of the direct feedback method, an iterative technique is applied to find the appropriate voltage input for the actuator that results in synchronous motion between the physical and virtual substructures and is free from the feedback-related instabilities.

Keywords Dynamic substructuring · Hardware-in-the-loop · Hybrid testing

7.1 Introduction

Hybrid experiments, alternatively known as hardware-in-the-loop (HIL), are where the tested system is divided into separate physical and numerical substructures that interact with each other through sensors and actuators [1]. A hybrid experiment can be useful in many circumstances, e.g. it allows for easier testing of components of prototypes if only certain parts of a structure need to be manufactured or, by making it possible to test the physical substructure with different numerical counterparts, it can be used as an aid in deciding on the final design.

Conducting such experiments however may be cumbersome. The contribution of the interface (sensors and actuators) to the dynamics of the overall system may cause stability and/or fidelity issues and so the hybrid test may show a different behaviour from the true assembly.

The traditional approach to overcome these challenges is to use an interface controller [2] or inverse model compensation [3] when generating the input to the physical substructure. However, these require either robust control design or accurate modelling of all the sensors and actuators used in the experiment, both of which can be difficult to obtain and must be done repeatedly if the experimental set-up is changed.

We propose an iterative method for hybrid testing for the case where steady-state responses are of interest and demonstrate the method on a benchtop-scale three-storey building. By breaking the feedback loop, the iterative method eliminates the potential stability and fidelity issues the control or inverse model-based approach may involve.

7.2 Measurement Set-Up

In our study, we perform the dynamic substructuring on a benchtop-scale three-storey building. The experimental set-up is shown in Fig. 7.1. In the hybrid experiment, the top two storeys are considered as the physical substructure and are connected to a real-time simulation of a virtual storey at the bottom of the building (see the 'True assembly' in Fig. 7.1). The bottom storey of the physical building, connected to an electrodynamic shaker, is used as the interface between the physical and virtual substructures. The force between the interface and the physical substructure is measured by a load cell and fed back

S. Beregi (✉) · D. A. W. Barton · D. Rezgui · S. A. Neild
Faculty of Engineering, University of Bristol, Bristol, UK
e-mail: sandor.beregi@bristol.ac.uk; david.barton@bristol.ac.uk; djamel.rezgui@bristol.ac.uk; simon.neild@bristol.ac.uk

M. Allen et al. (eds.), *Dynamic Substructures, Volume 4*, Conference Proceedings of the Society for Experimental Mechanics Series,
https://doi.org/10.1007/978-3-031-04094-8_7

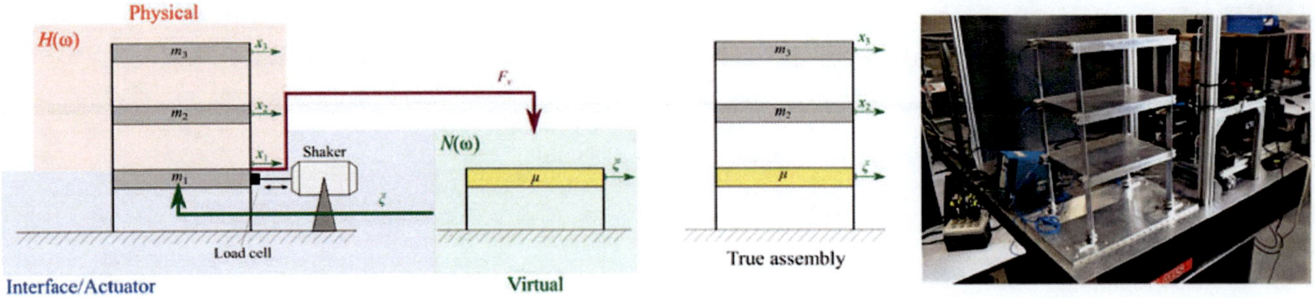

Fig. 7.1 Left panel: Dynamic substructuring of the benchtop three-storey building. Right panel: Experimental rig with the benchtop-size three-storey building

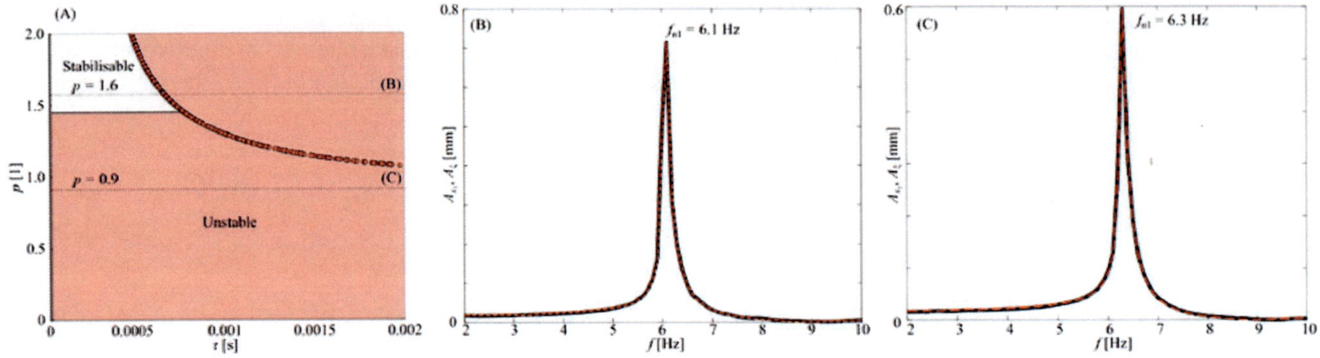

Fig. 7.2 Panel (A): Stabilisability diagram of the HIL experiment with direct displacement feedback from the virtual substructure in the plane of the feedback time-delay τ and the mass ratio p of the virtual storey and the interface. Panels (B) and (C): Frequency response of the hybrid system measured with the iterative technique. The amplitudes of the interface are indicated by the thick black continuous curves, while the dashed orange curves correspond to the amplitude of the virtual storey displacement

to the simulation of the virtual bottom storey, whereas we use the shaker to synchronise the displacement of the interface with the virtual substructure.

7.3 Substructurability Analysis

The main challenge in hardware-in-the-loop experiments is that the sensors and the actuators also contribute to the overall dynamics of the hybrid system which may lead to instability or fidelity issues in the experiment. In the case of the three-storey building experiment, the impact of the electrodynamic shaker on the hybrid system is more prominent compared to the sensors. Therefore, we focus on the effect of actuator dynamics in the assessment of the system's substructurability [1] with a feedback-based approach, that is, the ease by which a hybrid test can be conducted.

The contribution of the shaker's internal dynamics is perceived as a phase lag between the input voltage and the output force. This lag is modelled as a time-delay in the displacement feedback yielding to delay differential equations (DDE) as equations of motion [4]. The stability analysis of the DDE system (see panel (A) in Fig. 7.2.) reveals that the largest delay for which the feedback control is stabilisable is very small ($\tau_{crit} < 10^{-3}$ s). Moreover, there is a lower limit for the mass ratio $p = \mu/m_1$ of the virtual storey and the interface. This means that feedback control-based substructuring is impossible if a small mass is considered in the virtual substructure, but the task is also very challenging in the theoretically feasible parameter domains.

7.4 Iterative Method for Hybrid Testing

To avoid the issues of the feedback-based approach, an iterative method is used to achieve the synchronous motion of the interface and the virtual storey. While this method restricts us to tracking only steady-state periodic solutions of the hybrid system, it arguably provides higher-quality data in these scenarios, and such data still covers many common applications in engineering, e.g. measuring the frequency response of the system. By assuming that the experiment is adequately described by a linear system, the displacement of the virtual storey and the interface is represented by sinusoids. Then we use Broyden's method to find the phase and amplitude of the voltage signal, sent to the actuator, that results in a synchronous motion between the virtual and the physical substructures.

In panels (B) and (C) of Fig. 7.2, two frequency responses of the physical-virtual hybrid system, measured using the iterative technique, are presented. The two response curves in each diagram indicate very good agreement between the demand and the achieved interface displacement. These results demonstrate that this method can be used with equal success irrespective of whether the parameters are selected from the domain where direct displacement feedback is feasible (see the lines in panel (A) of Fig. 7.1 indicating the mass ratios corresponding to cases (B) and (C)).

It is worth mentioning the similarity of the iterative technique and control-based continuation [5], which can be used to track limit cycles in physical nonlinear systems. In fact, even though we only measure stable solutions in this experiment, one can use displacement control on the interface to reduce long transients in systems with low damping, as is the case with the three-storey structure. Thus, instead of the voltage signal sent to the actuator, the harmonic coefficients of the control target are considered as independent variables in the iteration. It is worth to point out that this does not mean direct feedback from the virtual substructure, since the synchronisation of the displacements of the interface and the virtual storey is still achieved iteratively. Therefore, while the transients are significantly reduced, allowing for a faster convergence of the iteration, the steady-state response is unaffected.

7.5 Conclusions

This study demonstrated that breaking the feedback loop and using an iterative method can be used for the dynamic substructuring of a benchtop-size three-storey building under periodic excitation in scenarios when hybrid testing is challenging or even infeasible with standard techniques. The main advantage of the proposed method is that it is free from the instabilities related to the feedback between the virtual and physical substructures. While the iterative technique does not require direct feedback control, a displacement control, where the target is set iteratively, can be used to improve its resilience to the fluctuations caused by transients enabling faster measurement of frequency responses in weakly damped systems.

The iterative method also overcomes the main challenge arising in techniques where the feedback from the virtual storey is realised through an inverse model of the interface, namely, this approach requires an excessive identification procedure which should be repeated every time a significant change is made to the experimental set-up. Moreover, one may find some parts of the underlying dynamics (e.g. nonlinearities) difficult to capture. In contrast, the proposed iterative technique can be used even with little or no knowledge about the actuator dynamics; therefore, it can be a robust and versatile tool for hardware-in-the-loop experiments.

Acknowledgements This research has received funding from the Digital twins for improved dynamic design (EP/R006768/1) EPSRC grant. The support of the EPSRC is greatly acknowledged.

References

1. Terkovics, N., Neild, S.A., Lowenberg, M., Szalai, R., Krauskopf, B.: Substructurability: the effect of interface location on a real-time dynamic substructuring test. Proc. R. Soc. A. **472**, 20160433 (2016)
2. Wallace, M.I., Wagg, D.J., Neild, S.A.: An adaptive polynomial based forward prediction algorithm for multi-actuator real-time dynamic substructuring. Proc. R. Soc. A. **461**, 3807–3826 (2005)
3. Ruffini, V., Szczygolowski, C., Barton, D.A.W., Lowenberg, M., Neild, S.A.: Real-time hybrid testing of strut-braced wing under aerodynamic loading using an electrodynamic actuator. Exp. Tech. **44**, 821–835 (2020)
4. Stepan, G.: Retarded Dynamical Systems: Stability and Characteristic Functions. Longman Scientific & Technical, Essex (1989)
5. Barton, D.A.W., Burrow, S.: Numerical continuation in a physical experiment: investigation of a nonlinear energy harvester. J. Comput. Nonlinear Dyn. **6**, 011010 (2011)

Chapter 8
Accuracy of Nonlinear Substructuring Technique in the Modal Domain

Jacopo Brunetti, Walter D'Ambrogio, Annalisa Fregolent, and Francesco Latini

Abstract The dynamic analysis of complex engineering systems can be performed by considering the assemblies as composed of subsystems and coupling them through substructuring techniques. A technique in the modal domain called Nonlinear Coupling Procedure (NLCP) has been recently defined to couple subsystems connected through nonlinear connections by using their Nonlinear Normal Modes (NNMs). It manages to capture the main dynamic features of the system, i.e., the backbone of each NNM, and it provides satisfactory results in terms of mode shape and resonance frequency as function of the excitation level of the assembled system, with a considerable reduction of the computational time. However the results may be inaccurate due to approximations either in the models of the subsystems or due to the considered coupling technique. Furthermore, nonlinear subsystems can be the cause of complex behaviors of the assembly and then their models need to be carefully characterized. Thus, it is necessary to evaluate the reliability of the method in terms of the accuracy of the solution. This is done by defining a reliability ratio based on energy concepts depending on the level of the excitation acting on the system. The effectiveness of the reliability ratio of nonlinear techniques is verified on the NLCP applied to a mechanical system with localized nonlinearities.

Keywords Nonlinear coupling · Dynamic substructuring · Nonlinear normal modes · Nonlinear connection · Reliability ratio

8.1 Introduction

In the recent years, many techniques have been developed to perform dynamic analysis of complex systems in order to determine whether the assemblies could correctly work in their operating conditions. Dynamic substructuring techniques have been broadly used in order to perform dynamic analysis of complex systems composed by multiple parts connected together [1]. The analysis can be performed in the frequency, modal, and time domain and it is possible to perform coupling and decoupling analysis combining different models, either numerical and experimental [2]. Using these techniques, each component can be modeled separately and the behavior of the complete system can be obtained more easily than if it was studied as a whole. The coupling problem [3, 4] is fundamental when a new part needs to be coupled to an existing assembly and it is necessary to evaluate how it could affect the dynamics of the new system. The decoupling [5, 6] is performed when the dynamics of a component needs to be extracted from the one of the assembly. For example, Saeed et al. [7, 8] used the System Equivalent Model Mixing (SEMM) technique together with the Virtual Point (VP) transformation to extract the property of the joint between a blade and a turbine rotor.

However, these techniques are all based on the assumption that the systems are linear. This can be a limit, especially when it is known that the systems experience nonlinear phenomena in their operating conditions. For those reasons, many techniques have been developed over the years in order to account for nonlinear effects using substructuring procedures. Kalaycioğlu and Özgüven developed an approach in the frequency domain to perform coupling [9, 10] and decoupling [11] procedures when one of the subsystems is nonlinear. Note that the presence of nonlinearity is taken into account by using

J. Brunetti · W. D'Ambrogio
Dipartimento di Ingegneria Industriale e dell'Informazione e di Economia, Università dell'Aquila, L'Aquila, Italy
e-mail: jacopo.brunetti@univaq.it; walter.dambrogio@univaq.it

A. Fregolent (✉) · F. Latini
Dipartimento di Ingegneria Meccanica e Aerospaziale, Università di Roma La Sapienza, Rome, Italy
e-mail: annalisa.fregolent@uniroma1.it; francesco.latini@uniroma1.it

M. Allen et al. (eds.), *Dynamic Substructures, Volume 4*, Conference Proceedings of the Society for Experimental Mechanics Series,
https://doi.org/10.1007/978-3-031-04094-8_8

Describing Functions (DFs) [12]. Kuether et al. [13] proposed instead a method based on the Implicit Condensation and Expansion procedure (ICE) [14] in order to obtain a Nonlinear Reduced Order Model (NLROM) of the system.

Another method is described in [15] and proposed by Kuether and Allen, in which the nonlinear substructures are coupled together in the modal domain by using Nonlinear Normal Modes (NNMs). This method was later improved by the authors in [16, 17] by considering the connections themselves as substructures [18] and by increasing the rate of convergence of the whole procedure. This technique, called Nonlinear Coupling Procedure (NLCP), manages to capture the main dynamics of the system and it provides satisfactory results in terms of mode shape and resonance frequency at given energy levels, with a considerable reduction of the computational time. Using this technique it is possible to analyze the effects of hardening and softening connections [19] and show the effectiveness of the method when different types nonlinearities are present [20]. However, the results may be inaccurate due to approximations either in the models of the subsystems or due to the considered coupling technique. In fact, the NLCP is based on a quasi-linear approximation and, as shown in [16, 18], it manages to retrace the main backbone of the NNMs of the coupled assembly, but neglects the contribution due to sub- and super-harmonics, which might become relevant during the nonlinear oscillations. As a result, the NNMs obtained using the NLCP may not be compliant with the definition given by Kerschen et al. in [21], because the response may be slightly a-periodic. For this reason, it can be useful to evaluate the reliability of the method in terms of the accuracy of the solution. In this paper, this is done by defining a reliability ratio based on energy concepts that evaluate the contribution of the main harmonic and its super-harmonics in the response of the system at each excitation level. The effectiveness of the reliability ratio of nonlinear techniques is verified on the NLCP applied to a mechanical system composed of two beams jointed together through two nonlinear connecting elements.

8.2 Nonlinear Coupling Procedure in Modal Domain

The Nonlinear Coupling Procedure described in [17] is a substructuring technique set in the modal domain that manages to retrace the backbone of the NNMs of undamped nonlinear systems. In particular, the attention is focused on all those cases in which two or more subsystems are connected together through nonlinear connections, such as wire rope isolators and bolted joints. In order to use this method, the nonlinear connection needs to be included in the substructuring process as a substructure itself by using its NNMs. This allows to characterize each substructure separately and then to couple them, thus reducing the overall computational burden. In fact, in this case the NNMs of the nonlinear substructures (i.e., nonlinear connections) are computed for a simple model (a portion of the whole mechanical system), and linear substructures can be reduced using a truncated subset of their Linear Normal Modes (LNMs). However, the present technique is based on the assumption that the nonlinear forces are harmonic at the same frequency of the excitation, thus sub- and super-harmonics effects are not detected. In fact, in most of the cases it is sufficient to consider the contribution of the main harmonic to obtain significant information regarding the dynamics of the main system. Sometimes, instead, other harmonics effects are needed to describe particular situations, thus alternative methods should be used, such as the pseudo-arclength continuation [22] or the Harmonic Balance (HB) [23]. Nevertheless, these methods are generally applied directly on the whole mechanical system, thus the computational burden could be non-negligible. The technique described in [17] allows to obtain the NNMs describing the variation of the resonance frequency and the mode shape of a nonlinear coupled system as function of the total energy stored in it. The NLCP performs the coupling in the modal domain using the reduction matrix $[R]$, defined as

$$[R]_{k,i} = \text{diag}([\Phi]^L, \left[\Psi(E_{k,i}^{NL})\right]^{NL}, \ldots) \tag{8.1}$$

where $[\Phi]^L$ is the truncated set of LNMs for the linear substructure and $[\Psi]^{NL}$ are the NNMs of the nonlinear connection evaluated at the energy E. It is based on an iterative algorithm of prediction-correction through which the main backbone of each NNM is retraced, as shown in Fig. 8.1. In particular, the procedure to evaluate the NNM at the k-th energy level is initiated using a predictor based on the energy distribution among substructures obtained at the previous energy levels. Then, the NNM is corrected by modifying for each iteration i the energy level $E_{k,i}$ of the NNMs $[\Psi]^{NL}$ included in $\left[R_{k,i}\right]$, until convergence is achieved.

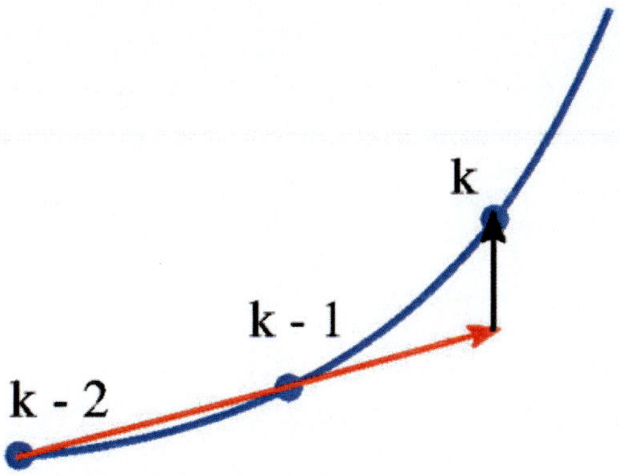

Fig. 8.1 Continuation procedure: the blue solid line is the NNM to be identified, the red arrow indicates the prediction step, the black one refers to the correction step, and the blue dots are the identified points of the NNM

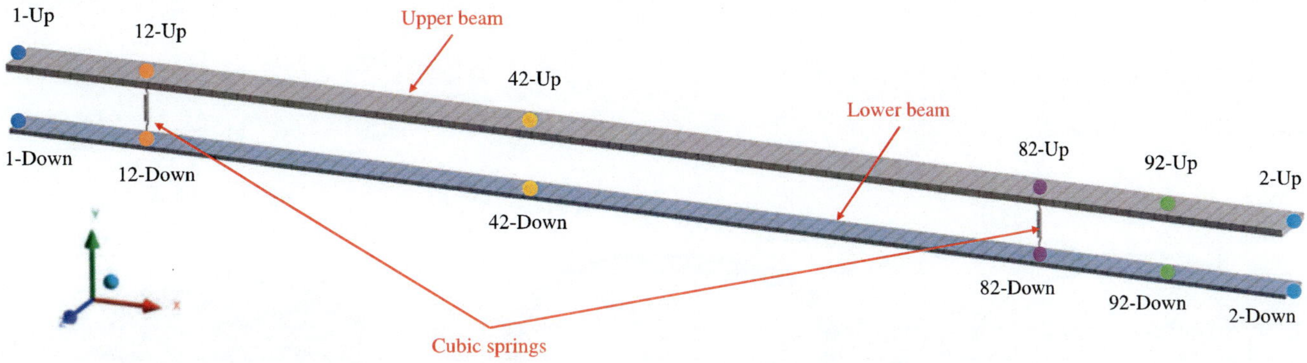

Fig. 8.2 The system is composed of one thick beam (upper one), one thin beam (lower one), and two cubic springs; the colored dots represent the nodes retained for the Craig-Bampton reduction of each beam

Table 8.1 Values of the physical and geometrical parameters of the upper and lower beam

L_{up}	w_{up}	t_{up}	L_{low}	w_{low}	t_{low}	ρ	E	n_{up}	n_{low}
1000 mm	40 mm	5 mm	1000 mm	30 mm	3 mm	7850 kg/m^3	200 GPa	101	101

8.2.1 Numerical Model

The system under consideration is shown in Fig. 8.2, and its geometrical and mechanical properties are listed in Table 8.1. The system is composed of two linear substructures, the upper and lower beam, connected through two nonlinear substructures, i.e., the cubic spring in Fig. 8.2, whose stiffness is

$$k_{cub}(x) = k_l + k_{nl}x^2 \tag{8.2}$$

where $k_l = 13610$ N/m and $k_{nl} = 1.33 \cdot 10^{10}$ N/m^3. These values are inferred from the analysis of a component similar to the one used in [16, 17], whose law of stiffness is accurately described by Eq. (8.2). A 2D problem on the XY plane is considered, and the system is in free-free condition. In order to perform the coupling, a Craig-Bampton reduction is performed on each beam by retaining only the y-displacement and the z-rotation of six nodes (colored dots in Fig. 8.2) and ten fixed interface modes.

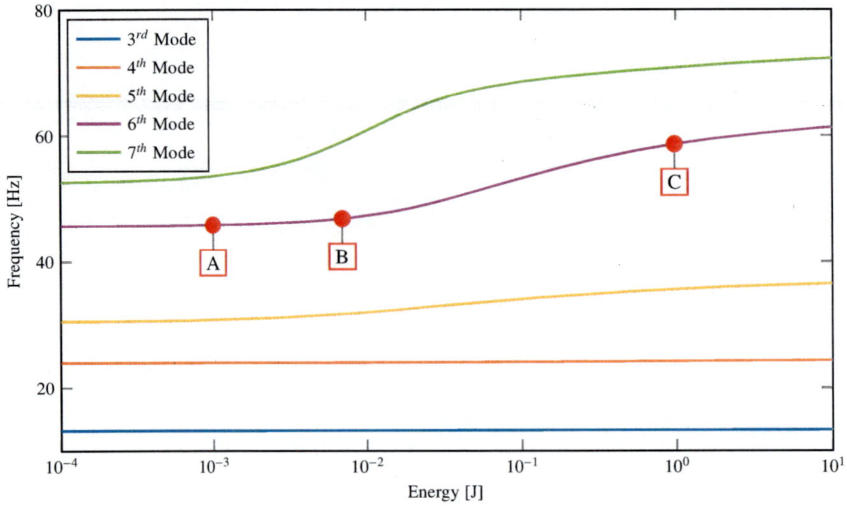

Fig. 8.3 Backbone of the first five NNMs of the system

Table 8.2 Computational time for the computed NNMs

Mode	3rd	4th	5th	6th	7th
Time [s]	5.3	5.6	6.5	10.7	7.6

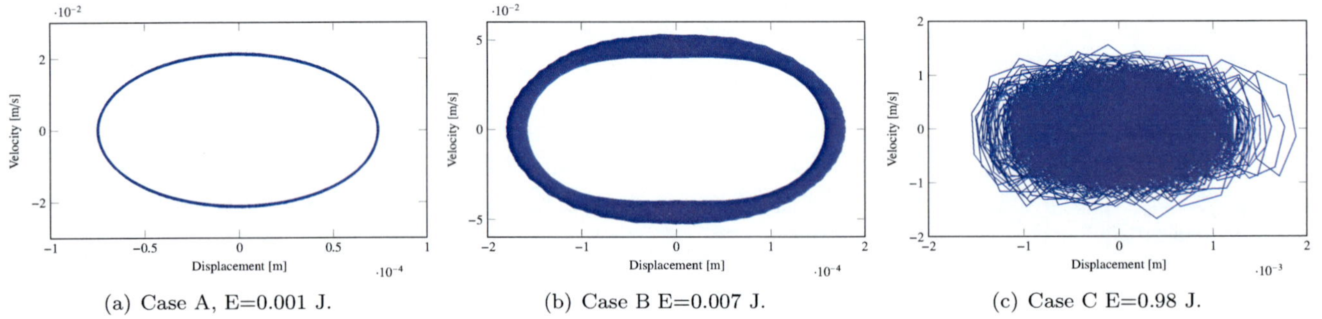

Fig. 8.4 State-space representation of the y-DoF of node 12-up at three different energy levels. (**a**) Case A, E=0.001 J. (**b**) Case B E=0.007 J. (**c**) Case C E=0.98 J

8.2.2 Results

The NLCP is applied to couple the two beams and the two cubic springs shown in Fig. 8.2, and it provides the main backbone of the first five NNMs of the structure, shown in Fig. 8.3.

Note that since the whole structure is unconstrained, the first two modes of vibration are rigid body modes, thus the first five nonrigid and nonlinear normal modes are computed. The use of the Craig-Bampton reduced matrices allows to compute the solution in a short time, as summarized in Table 8.2. The NLCP provides the mode shape for a set of given energy levels, for each mode. Considering for example the sixth mode, it is possible to solve the set of differential equations describing the dynamics of the mechanical system using the mode shape for the energy levels identified by labels (A,B,C) in Fig. 8.3, that are associated with low, medium, and high energy, respectively. Thus, the time response of the retained nodes is obtained and, for example, Fig. 8.4 shows the state-space representation of node 12-Up for the y-DoF. The state-space representation shows that as the energy level increases, the orbit becomes more complex. In particular, case A at low energy is almost periodic, while case B at medium energy is already affected by the nonlinearities (the orbit is not elliptic) and even if the periodicity is lost, significant information about the mode shape can be still retrieved. However, in case C at high energy the response provided by the NLCP is chaotic and no information can be inferred in terms of mode shape, although the main resonance frequency is correctly detected [17].

8.3 Reliability Ratio

The results provided by the NLCP might be inaccurate to fully describe the dynamics of complex systems in certain scenarios, thus it is necessary to define a reliability ratio in order to determine the energy intervals where the solution can be considered as accurate. Before going into the detail of the ratio, it is useful to consider a linear structure of which the mode shapes and the correspondent resonance frequencies are known. Supposing to consider one of the modes of vibration with resonance frequency different from zero, the corresponding mode shape can be used as the initial condition to solve the set of differential equations and obtain its time response. Considering the time response of one of its DoFs, it is possible to perform a Power Spectral Density (PSD) and get its frequency content. Since the system is linear, it is straight-forward that the peak found in the PSD corresponds to the resonance frequency associated with the considered mode shape.

If the same procedure is performed when the nonlinear system is considered, a different result is attained. In fact, according to Kerschen's definition of a NNM [21], the response is still periodic and in the PSD there is not only the peak of the fundamental harmonic but also the ones of its sub- and super-harmonics. Since the NLCP does not account for the presence of sub- and super-harmonics, it can provide inaccurate results when the nonlinear effects become more evident, i.e., as the energy increases. In this case, the presence of harmonics different from the fundamental one is detected, but the power associated with them is not correctly distributed, resulting in the appearance of other spurious harmonics. As an example, in Fig. 8.5 the PSDs of the y-displacement of node 12-Up for the cases A,B, and C in Fig. 8.4 are shown.

It is evident from Fig. 8.5a, b that power is concentrated around the fundamental harmonic ($\approx$45 Hz) and the third super-harmonic ($\approx$135 Hz), confirming the good estimation of the mode shape used as initial conditions. Instead, the PSD in Fig. 8.5c highlights that there is not any relevant peak, thus the estimation of the mode shape is not correct as already pointed out in Fig. 8.4c. Therefore, in order to evaluate the energy range in which the solution obtained using the NLCP can be considered as accurate, a reliability ratio (R_r) is computed. It compares the power associated with the fundamental harmonic and its super-harmonics P_{harm} to the power associated with the whole spectrum P_{tot} as follows:

$$P_{harm,e} = \sum_{n=1}^{N} PSD_{e,n} \tag{8.3}$$

$$P_{tot,e} = \int_0^{f_{max}} PSD_e(f)df \tag{8.4}$$

$$R_r(e) = \frac{P_{harm,e}}{P_{tot,e}} \tag{8.5}$$

where e indicates the energy level at which the ratio is evaluated, N indicates the number of considered super-harmonics, and f_{max} is the maximum frequency at which the integral is computed. Each term in the sum in (8.3) is computed as the integral of the PSD around the frequency nf_{res}, being f_{res} the fundamental frequency. If this ratio is close to one, it indicates that the power is correctly distributed among the harmonics and there is not any loss on other spurious harmonics. If it decreases, instead, it means that spurious harmonics are present and the mode shape result may not be considered as acceptable. For example, for the cases shown in Fig. 8.5, the reliability ratio is equal to $R_{r,A} = 0.999$, $R_{r,B} = 0.985$, and $R_{r,C} = 0.202$.

Therefore, once the NNMs are computed, it is possible to evaluate the reliability ratio on some nodes that are considered as meaningful in the dynamics of the system. In this case, the attention is focused on the nodes at the end of the nonlinear

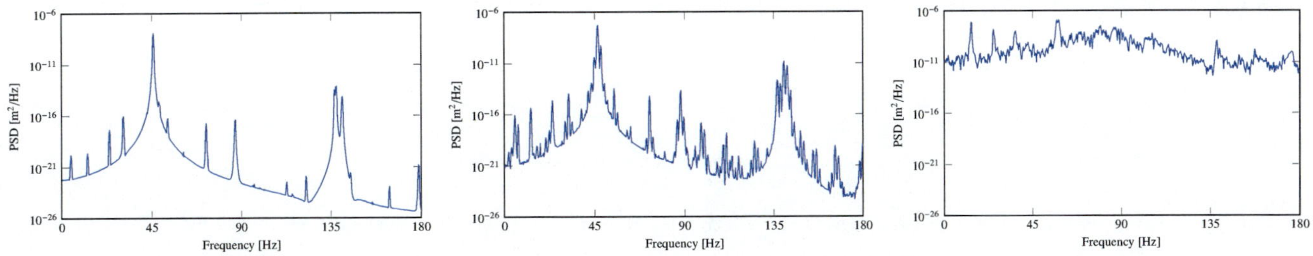

(a) Case A. (b) Case B. (c) Case C.

Fig. 8.5 PSDs of the y-displacement response of node 12-up for cases A, B, and C. (**a**) Case A. (**b**) Case B. (**c**) Case C

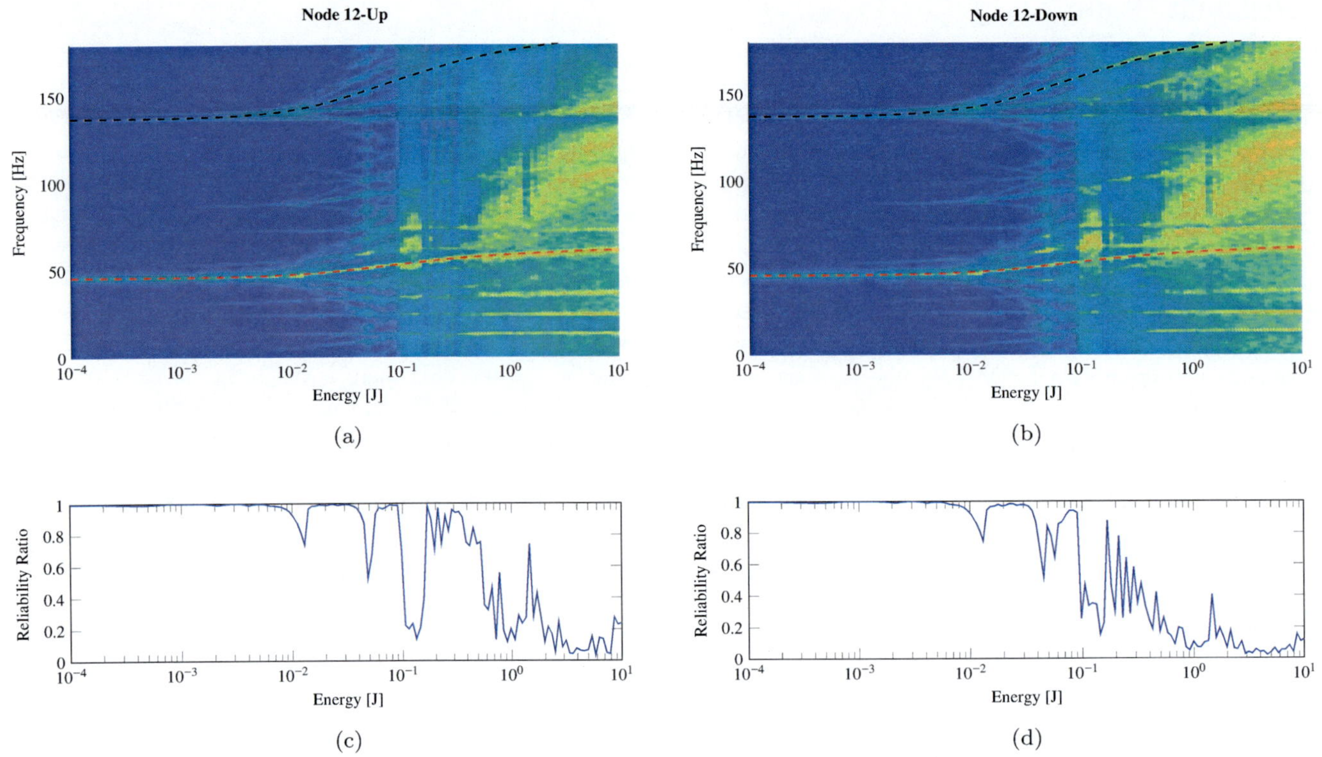

Fig. 8.6 Results for nodes 12-up and 12-down. (**a**) Collection of PSDs at different energy levels; (red dashed line) fundamental harmonic, (black dashed line) third super harmonic. (**b**) Collection of PSDs at different energy levels; (red dashed line) fundamental harmonic, (black dashed line) third super harmonic. (**c**) Accuracy ratio evaluated for node 12-up. (**d**) Accuracy ratio evaluated for node 12-down

springs, since these are the sources of nonlinearity in the structure: node 12-Up, node 12-Down, node 82-Up, and node 82-Down from Fig. 8.2. The same procedure performed to obtain the PSDs for cases A, B, and C shown in Fig. 8.5 can be generally extended to all the energy levels at which the NNM is obtained. The results can be gathered in a single graph as the one shown in Fig. 8.6a for node 12-Up, where each vertical line corresponds to a colored scale representation of the power spectral density for the given energy level. This provides a graphical representation of how the frequency content of the response varies as the energy increases.

The results for nodes 12-Up and 12-Down are shown in Fig. 8.6. From Fig. 8.6a, b it is possible to see that up until E= 0.1 J the vibration power is basically concentrated on the main harmonic (red dashed line at ≈45 Hz) and the third super-harmonic (black dashed line at ≈135 Hz). Increasing the energy, the power is spread among all the considered frequency bandwidth. Figure 8.6c, d show the reliability ratio for the PSDs shown in Fig. 8.6a, b, respectively. The reliability ratios share a similar trend of variation as function of the energy and are close to one up until E= 0.1 J while they are consistently below one beyond E= 0.01 J, confirming what already inferred from the visual analysis of the PSDs. A similar result can be obtained for the nodes placed at the end of the other cubic springs (Fig. 8.7), node 82-Up and node 82-Down, meaning that the reliability ratio does not depend on the node at which it is calculated but it is associated with the dynamics of the complete system.

This ratio is an important indicator because it defines the level of accuracy of the solution as function of the energy level at which it is evaluated. It can be computed very easily and provided along with the solution of the NLCP (or any other method) in order to assess the validity of the obtained results.

8.4 Conclusions

The substructuring procedure in the modal domain called NLCP is used to couple two beams connected through two nonlinear springs. The effectiveness of the technique is evaluated through the accuracy of the results by defining a reliability ratio. The time response of the DoFs of the structure is computed and the PSD is performed in order to evaluate their frequency content. Then, for each energy level, the ratio between the power associated with the main harmonic and its super-

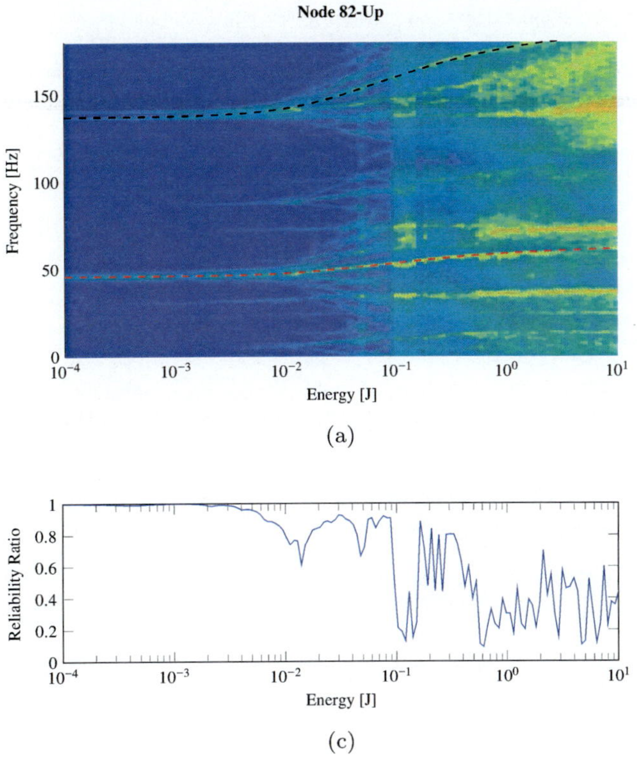

Fig. 8.7 Results for nodes 82-up and 82-down. (**a**) Collection of PSDs at different energy levels; (red dashed line) fundamental harmonic, (black dashed line) third super harmonic. (**b**) Collection of PSDs at different energy levels; (red dashed line) fundamental harmonic, (black dashed line) third super harmonic. (**c**) Accuracy ratio evaluated for node 82-up. (**d**) Accuracy ratio evaluated for node 82-down

harmonics and the power of the whole spectrum is computed. In this way it is possible to understand how much power loss there is in the computed response and evaluate the energy range in which the procedure is reliable. From the numerical data obtained by coupling the two beams, it seems that the procedure can be considered as reliable until the energy content of the system reaches 0.1 J. At higher energy levels, the ratio is far below one, meaning that the power is spread among the whole frequency bandwidth and the results in terms of mode shape cannot be considered as acceptable. Thus, if the ratio is almost one the solution can be considered as accurate in the given energy interval, whereas if it approaches zero it means that other methods should be used. It is important to notice that the proposed reliability ratio can be computed starting from the results of any nonlinear technique. Therefore it is a valid indicator to evaluate the reliability of all the techniques that deal with the dynamics of complex structures.

Acknowledgments This research is supported by University of Rome La Sapienza and University of L'Aquila.

References

1. de Klerk, D., Rixen, D.J., Voormeeren, S.N.: General framework for dynamic substructuring: history, review, and classification of techniques. AIAA J. **46**(5), 1169–1181 (2008)
2. Brunetti, J., Culla, A., D'Ambrogio, W., Fregolent, A.: Experimental dynamic substructuring of the Ampair wind turbine test bed. In: Conference Proceedings of the Society for Experimental Mechanics Series, vol. 1, pp. 15–26. Springer New York LLC (2014)
3. Craig, R. Jr.: Coupling of substructures for dynamic analyses-an overview. In: 41st Structures, Structural Dynamics, and Materials Conference and Exhibit, p. 1573 (2000)
4. Rixen, D.J.: A dual Craig-Bampton method for dynamic substructuring. J. Comput. Appl. Math. 168(1–2), 383–391 (2004)
5. Voormeeren, S.N., Rixen, D.J., A family of substructure decoupling techniques based on a dual assembly approach. Mech. Syst. Signal Process. **27**, 379–396 (2012)
6. D'Ambrogio, W., Fregolent, A., Decoupling procedures in the general framework of frequency based substructuring. In Proceedings of 27th IMAC. Orlando (USA) (2009)

7. Saeed, Z., Klaassen, S.W.B., Firrone, C.M., Berruti, T.M., Rixen, D.J.: Experimental joint identification using system equivalent model mixing in a bladed-disk. J. Vib. Acoust. **142**(5), 1–29 (2020)
8. Saeed, Z., Firrone, C.M., Berruti, T.M.: Joint identification through hybrid models improved by correlations. J. Sound Vib. **494**, 115889 (2021)
9. Kalaycioğlu, T., Özgüven, H.N.: Harmonic response of large engineering structures with nonlinear modifications. In Proceedings of the 8th International Conference on Structural Dynamics, EURODYN, pp. 3623–3629 (2011)
10. Kalaycıoğlu, T., Özgüven, H.N.: Nonlinear structural modification and nonlinear coupling. Mech. Syst. Signal Process. **46**(2), 289–306 (2014)
11. Kalaycıoğlu, T., Özgüven, H.N., FRF decoupling of nonlinear systems. Mech. Syst. Signal Process. **102**, 230–244 (2018)
12. Vander Velde, W.E., Multiple-Input Describing Functions and Nonlinear System Design. McGraw-Hill, New York (1968)
13. Kuether, R.J., Allen, M.S., Hollkamp, J.J.: Modal substructuring of geometrically nonlinear finite-element models. AIAA J. **54**(2), 691–702 (2015)
14. Hollkamp, J.J., Gordon, R.W., Spottswood, S.M.: Nonlinear modal models for sonic fatigue response prediction: a comparison of methods. Shock Vib. Digest **38**(3), 232–233 (2006)
15. Kuether, R.J., Allen, M.S., Nonlinear modal substructuring of systems with geometric nonlinearities. In 54th AIAA/ASME/ASCE/AHS/ASC Structures, Structural Dynamics, and Materials Conference, p. 1521 (2013)
16. Latini, F., Brunetti, J., Kwarta, M., Allen, M.S., D'Ambrogio, W., Fregolent, A.: Experimental results of nonlinear structure coupled through nonlinear connecting elements. In Proceedings of ISMA 2020 - International Conference on Noise and Vibration Engineering and USD 2020 - International Conference on Uncertainty in Structural Dynamics (2020)
17. Latini, F., Brunetti, J., D'Ambrogio, W., Allen, M.S., Fregolent, A.: Nonlinear substructuring in the modal domain: numerical validation and experimental verification in presence of localized nonlinearities. Nonlinear Dyn. **104**(2), 1043–1067 (2021)
18. Latini, F., Brunetti, J., D'Ambrogio, W., Fregolent, A.: Substructures' coupling with nonlinear connecting elements. Nonlinear Dyn. **99**(2), 1643–1658 (2020)
19. Brunetti, J., D'Ambrogio, W., Fregolent, A., Latini, F.: Substructuring using NNMs of nonlinear connecting elements. In: Carcaterra, A., Paolone, A., Graziani, G. (eds.) Proceedings of XXIV AIMETA Conference 2019. AIMETA 2019. Lecture Notes in Mechanical Engineering. Springer (2020)
20. Brunetti, J., D'Ambrogio, W., Fregolent, A., Latini, F.: Dynamic substructuring using a combination of softening and hardening connecting elements. In Dynamic Substructures, vol. 4, pp. 23–33. Springer, Berlin (2022)
21. Kerschen, G., Peeters, M., Golinval, J., Vakakis, A.F.: Nonlinear normal modes, part i: a useful framework for the structural dynamicist. Mech. Syst. Signal Process. **23**(1), 170–194 (2009)
22. Peeters, M., Viguié, R., Sérandour, G., Kerschen, G., Golinval, J.: Nonlinear normal modes, part ii: toward a practical computation using numerical continuation techniques. Mech. Syst. Signal Process. **23**(1), 195–216 (2009)
23. Krack, M., Gross, J.: Harmonic Balance for Nonlinear Vibration Problems. Springer, Berlin (2019)

Chapter 9
Quantification of Bias Errors Influence in Frequency Based Substructuring Using Sensitivity Analysis

Gregor Čepon, Domen Ocepek, Jure Korbar, Tomaž Bregar, and Miha Boltežar

Abstract The methodology to divide complex systems into several subsystems is a common practice in the field of structural dynamics. Structural dynamic analyses can be carried out more efficiently if subsystems are analysed separately and later coupled using dynamic substructuring techniques. However a reliable experimental application of frequency based substructuring (FBS) remains a challenge as it requires highly accurate acquisition of subsystems' frequency response functions (FRFs). Even a small error from the measurement campaign can yield erroneous coupling results. The measurement errors can be either random or systematic, with the latter often referred to as bias. Impact excitation is popular in dynamic substructuring due to the fast FRF calculation for each impact location. However, deviations in the location of excitation as a typical example of measurement bias affect the FRFs throughout the whole frequency range.

This paper proposes a novel methodology to characterize bias errors in FBS based on the small deviations in impact excitation from typical experimental measurements. The deviations are utilized to reconstruct a range of bias-affected FRFs. These are then used in the global sensitivity analysis in order to characterize how each impact location affects an arbitrary quality indicator, such as FRFs reciprocity or passivity. Therefore, the effect of bias can be evaluated directly from a single measurement campaign, without the need for a numerical model. The proposed approach is shown on a synthetic numerical example, where the advantages and limitations are outlined. Virtual point transformation is applied to obtain admittance matrix at the substructures' interfaces. Hence, virtual point reciprocity is proposed as a criterion to quantify bias error influence. Finally, an application of frequency based substructuring on beam-like structure is depicted.

Keywords Frequency based substructuring · Bias error · Virtual point transformation · Global sensitivity analysis

9.1 Introduction

Frequency based substructuring methods [1] are usually related to the experimental approach, as it is possible to define the exact dynamic properties directly based on the measured frequency response functions. Although FBS methods are well established, the challenge to provide accurate and reliable dynamic properties for the individual subsystem remains. Accurate measurement of the FRFs is essential, as the errors in the substructure's FRFs are propagated in the assembled FRFs [2]. In general, the measurement errors can be classified according to their nature into two categories: random errors and systematic (bias) errors. An error defined as systematic is an error with a consistent and repeatable nature. The effect of bias can be observed as a systematic shift in the measurement results; therefore, it does not affect the reliability but rather the accuracy of the outcome.

Here, a sensitivity-based approach is proposed to characterize the bias errors in FBS. Sensitivity analysis is introduced to characterize the influence of the impact location offset with respect to the consistency of the virtual point (VP), which is applied to couple both substructures. As the virtual point transformation (VPT) is based on the relative locations and orientations of the impacts and responses with respect to the VP, bias errors affect the VP admittance. In order to perform the global sensitivity analysis exclusively with the experimental response model, numerous biased FRFs should be measured

G. Čepon (✉)
University of Ljubljana, Faculty of Mechanical Engineering, Ljubljana, Slovenia

Gorenje d.o.o., Velenje, Slovenia

D. Ocepek · J. Korbar · T. Bregar · M. Boltežar
University of Ljubljana, Faculty of Mechanical Engineering, Ljubljana, Slovenia

M. Allen et al. (eds.), *Dynamic Substructures, Volume 4*, Conference Proceedings of the Society for Experimental Mechanics Series,
https://doi.org/10.1007/978-3-031-04094-8_9

at each impact location. This is practically impossible. Therefore, a linear relationship between the biased FRFs and the location offset was adopted [3] in order to be able to construct a sizable FRF set following Saltelli sampling scheme [4]. FRF set is intended to use in Sobol's sensitivity analysis [5] in order to simultaneously evaluate the relative contribution of each individual biased force input as well as their interactions on the variance of the reconstructed VP's reciprocity. The proposed method enables the identification of the impact locations where the bias error would have the greatest influence on the consistency of the VP. To present the efficiency of the proposed approach, a numerical case study is presented.

9.1.1 *Lagrange Multiplier Frequency Based Substructuring*

The LM FBS method makes it possible to determine the assembled system admittance $\mathbf{Y}^{\mathrm{AB}}$ from the subsystems FRFs. The equation of motion for the uncoupled substructures is

$$\boldsymbol{u} = \mathbf{Y}^{\mathrm{A|B}} \left(\boldsymbol{f} + \boldsymbol{g} \right). \tag{9.1}$$

The vector $\boldsymbol{u}$ represents the responses to the external forces $\boldsymbol{f}$ and $\boldsymbol{g}$ is the vector of interface forces between the substructures. Admittance matrices of each substructure are assembled into a block-diagonal matrix $\mathbf{Y}^{\mathrm{A|B}}$. The compatibility conditions between the substructures are written through the Boolean matrix $\mathbf{B}$ (Eq. (9.2)), which ensures that the substructures have the same displacements at the interface in the coupled state. The equilibrium conditions (Eq. (9.3)) are introduced using a set of unknown Lagrange multiplier vectors $\boldsymbol{\lambda}$.

$$\mathbf{B}\,\boldsymbol{u} = \mathbf{0}, \tag{9.2}$$

$$\boldsymbol{g} = -\mathbf{B}^{\mathrm{T}}\boldsymbol{\lambda}. \tag{9.3}$$

Solving the set of Eqs. (9.1)–(9.3) yields the response of the coupled structure:

$$\boldsymbol{u} = \mathbf{Y}^{\mathrm{AB}}\,\boldsymbol{f} = \left[\mathbf{Y}^{\mathrm{A|B}} - \mathbf{Y}^{\mathrm{A|B}}\,\mathbf{B}^{\mathrm{T}} \left(\mathbf{B}\,\mathbf{Y}^{\mathrm{A|B}}\,\mathbf{B}^{\mathrm{T}} \right)^{-1} \mathbf{B}\,\mathbf{Y}^{\mathrm{A|B}} \right] \boldsymbol{f}. \tag{9.4}$$

9.1.2 *Virtual Point Transformation*

A virtual point is chosen near the physical interface. Multiple responses and excitations are measured close to this point ($\mathbf{Y}_{\mathrm{uf}}$) and then projected onto the rigid interface deformation modes (IDMs). The transformation is achieved using the following equation:

$$\mathbf{Y}_{\mathrm{qm}} = \mathbf{T}_{\mathrm{u}}\mathbf{Y}_{\mathrm{uf}}\mathbf{T}_{\mathrm{f}}^{\mathrm{T}}, \tag{9.5}$$

where $\mathbf{T}_{\mathrm{u}}$ is the displacement transformation matrix and $\mathbf{T}_{\mathrm{f}}$ is the force transformation matrix. Both are assembled from the relative sensor/impact position and orientation [6]. $\mathbf{Y}_{\mathrm{qm}}$ is the VP admittance matrix with a perfectly collocated force/moment and translation/rotation DoF. Hence, the VP FRF matrix should be reciprocal. A reciprocity evaluation through coherence criterion can be used to assess the quality of the transformation:

$$\chi_{ij} = \mathrm{coh}\big(Y_{ij},\, Y_{ji}\big) = \frac{(Y_{ij} + Y_{ji})(Y_{ij}^{*} + Y_{ji}^{*})}{2\,(Y_{ij}\,Y_{ij}^{*} + Y_{ji}\,Y_{ji}^{*})}, \qquad Y_{ij},\, Y_{ji} \in \mathbf{Y}_{\mathrm{qm}}, \tag{9.6}$$

where $*$ denotes a complex conjugate. The criterion is bounded between 0 and 1, where the values closer to 1 indicate a strong correlation between two reciprocal VP FRFs.

9.2 Quantification of Bias Influence in FBS

An example of bias-affected FRFs for one impact location with multiple repetitions and one response location is shown in Fig. 9.1a. By plotting the bias-affected FRFs to the Nyquist plot at each frequency, a bias error is reflected in the FRFs distributed within an ellipse. Each impact, not aligning with the previous one, introduces new FRF. The ellipse's major axis is orientated in a direction where the FRFs are most sensitive to bias.

The equation of the ellipse's major axis can be determined using an approximation approach. First, the relation of the real FRF part with respect to the bias is determined using the minimum and maximum values:

$$\Re(\mathbf{Y}) = k_\mathrm{r}\,\mathbf{b} + n_\mathrm{r}. \tag{9.7}$$

The bias values $\mathbf{b}$ are bounded between -1 and 1. Next, the relation between the real and imaginary parts for the ellipse's major axis is obtained, fitting data from the complex plane to the following equation:

$$\Im(\mathbf{Y}) = k_\mathrm{i}\,\Re(\mathbf{Y}) + n_\mathrm{i}. \tag{9.8}$$

Using Eqs. (9.7) and (9.8) the FRFs for multiple biases in the most sensitive direction can be reconstructed. With the sizable FRF sample set available, each set of FRFs is then used in the VPT in order to obtain full-DoF FRFs at the VP. The coherence criterion (Eq. (9.6)) is used to assess the overall reciprocity of the VP FRF matrix (Fig. 9.1b). In the final step of the methodology, a Sobol sensitivity analysis is implemented on the set of reciprocity values from all the VP transformations according to [7]. With this approach, the bias on the impact locations that contribute significantly to the quality and repeatability of the VPT is recognized.

9.2.1 *Numerical Case Study*

Fifteen impact locations were positioned on a substructure B. For each location, multiple repetitions of impact excitation were performed, each with prescribed location offset. Accordingly to proposed approach, sensitivity analysis was carried out. The averaged sensitivity indices for the individual impact location are shown in Fig. 9.2.

Based on Fig. 9.2 the following sets of impacts are defined: 12 impacts with the lowest sensitivity indices are included in the LOW SENS set and 12 impacts with the highest sensitivity indices are incorporated in the HIGH SENS set (Table 9.1). By using the selection presented in Table 9.1 the B's VP FRFs (Fig. 9.3a) are constructed using the LOW SENS, HIGH SENS and all 15 biased impacts. Reference FRF is provided using a highly over-determined set of impacts and channels with no bias error. From Fig. 9.3a it is evident that by using the LOW SENS impact location set it is possible to increase the consistency of the VP transformation when compared to the reference.

In Fig. 9.3b the final coupling result for structures A and B is shown. At A, no bias errors were assumed when obtaining FRFs. Based on a visual inspection, it can be concluded that the coupled FRF based on LOW SENS impacts matches the reference with better agreement.

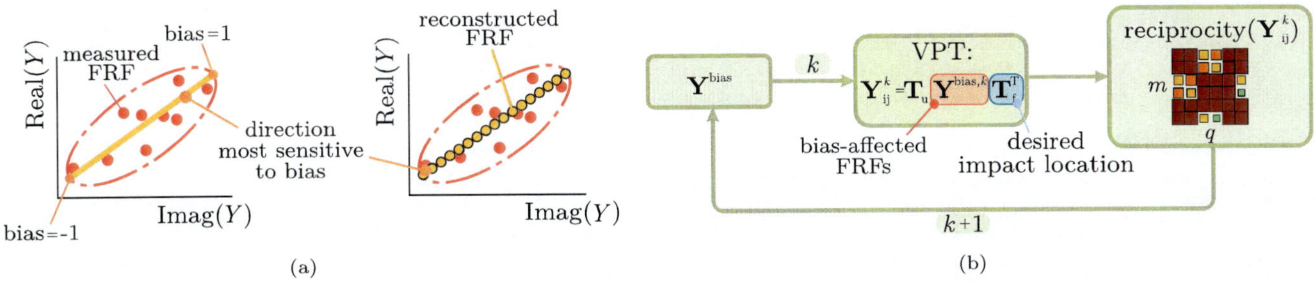

Fig. 9.1 (**a**) Reconstruction of multiple FRFs for each impact location using Saltelli sample scheme for set of biased excitation positions. (**b**) Estimation of VPT quality through the reciprocity criterion for each set of biased FRFs

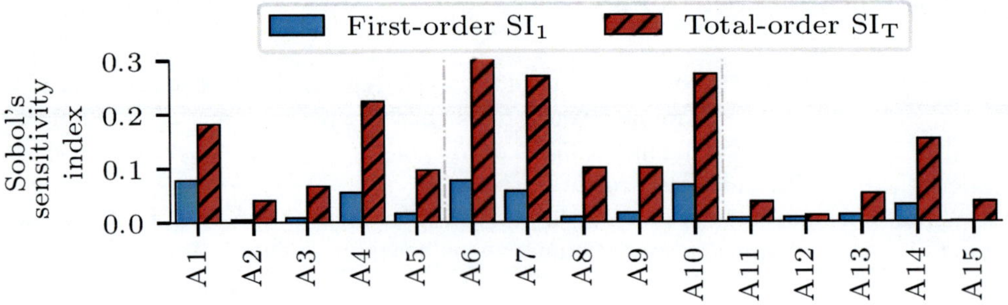

Fig. 9.2 First and total order Sobol's sensitivity indices for individual impact location.

Table 9.1 Impacts with low and high sensitivity to bias

LOW SENS	A2	A3	A4	A5	A7	A8	A9	A10	A11	A12	A13	A15
HIGH SENS	A1	A3	A4	A5	A6	A7	A9	A10	A11	A12	A13	A14

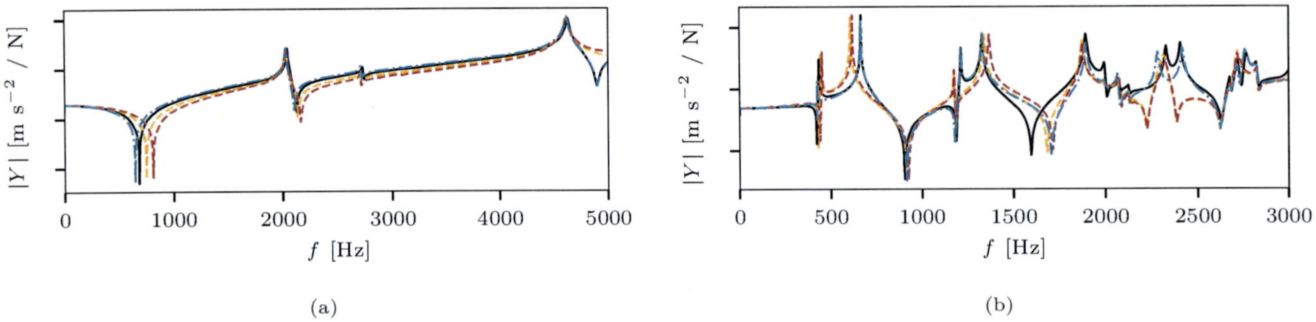

Fig. 9.3 (**a**) Comparison between B's VP FRFs for excitation in θ_y-direction and response in θ_y-direction using different sets of impacts. (**b**) Comparison between final assembly FRFs using different sets of impacts in process of obtaining $\mathbf{Y}^{\mathrm{B}}_{22}$. (black dashed line)—Reference, (blue dashed line)—LOW SENS, (red dashed line)—HIGH SENS, (orange dashed line)—all impacts

9.2.2 Conclusions

An approach to characterize the bias at the impact location was proposed. Characterization is performed directly from measurements without the need for an additional numerical model or dedicated experimental setup.

References

1. De Klerk, D., Rixen, D., Voormeeren, S.: General framework for dynamic substructuring: history, review and classification of techniques. AIAA J. **46**(5), 1169–1181 (2008)
2. Rixen, D.: How measurement inaccuracies induce spurious peaks in frequency based substructuring. In: Proceedings of the Twenty Sixth International Modal Analysis Conference, Orlando. Society for Experimental Mechanics, Bethel (2008)
3. De Klerk, D.: How bias errors affect experimental dynamic substructuring. In: Structural Dynamics, vol. 3, pp. 1101–1112. Berlin, Springer (2011)
4. A. Saltelli, Making best use of model evaluations to compute sensitivity indices. Comput. Phys. Commun. **145**(2), 280–297 (2002)
5. I.M. Sobol, Global sensitivity indices for nonlinear mathematical models and their Monte Carlo estimates. Math. Comput. Simul. **55**(1–3), 271–280 (2001)
6. van der Seijs, M., van den Bosch, D., Rixen, D., de Klerk, D.: An improved methodology for the virtual point transformation of measured frequency response functions in dynamic substructuring. COMPDYN (2013)
7. Bregar, T., Holeček, N., Čepon, G., Rixen, D., Boltežar, M.: Including directly measured rotations in the virtual point transformation. Mech. Syst. Signal Process. **141**, 106440 (2020)

Chapter 10
Real-Time and Pseudo-Dynamic Hybrid Simulation Methods: A Tutorial

Oh-Sung Kwon and Vasilis Dertimanis

Abstract This tutorial offers an introductory overview of hybrid simulation (HS) and its basic variants, i.e., real-time (RT) and pseudo-dynamic (PsD) HS. Starting from a conceptual representation of the HS loop and the establishment of a toy example, a few application examples are provided and the distinct differences between RTHS and PsDHS are noted. Accordingly, the fundamental attributes of the numerical substructure and the control plant (i.e., the transfer system and the experimental substructure) are explained, and two fundamental challenges of the HS loop, namely the compensation of the transfer system's dynamics and delay, are described. An adaptive, data-driven method for handling these challenges is then briefly presented.

Keywords Hybrid simulation · Dynamic substructuring · Real-time · Actuator control · Case studies

10.1 Introduction

The response of structural systems subjected to diverse loads is traditionally studied using numerical methods. Their accuracy inherently depends on many assumptions employed, either on the constitutive models or on the inelastic behavior of the individual structural components. In validating the behavior of numerical models, structural prototypes are usually tested in a laboratory. However, the laboratory tests are constrained by the space, the number of controllable degrees of freedoms, and actuators' force and stroke capacities. Further, the interaction between the specimen and the structural system cannot be properly modeled in the conventional structural element tests. Testing an entire structural system, such as a bridge or a building, is not impossible but very rare in the field of civil engineering.

To overcome the above limitations in the numerical and experimental methods, the hybrid (experimental–numerical) simulation methods have been actively investigated [1–3]. In a hybrid simulation (oftentimes referred to as *hardware-in-the-loop*), the elements that need to be represented with physical specimens are tested experimentally while the rest of the structural system is modeled numerically. The compatibility at the interface among the numerical and experimental substructures is achieved by controlling the deformations of the physical specimens based on the predicted displacements from the numerical model. Force equilibrium is achieved through a nonlinear solution algorithm in the numerical model. Implementing a hybrid simulation method requires a good understanding of numerical modeling, numerical integration schemes, substructuring, data communication, and control of actuators [4–6].

This tutorial covers an overview of the hybrid simulation methods and typical components required for carrying out real-time or pseudo-dynamic hybrid simulations. A few application examples will be provided. Then, an actuator control strategy will be presented for applications to real-time hybrid simulations. The tutorial session attendees will gain a good conceptual understanding of the hybrid simulation methods, main hardware components, and technical background required for implementing the hybrid simulations, as well as ideas for potential applications of the hybrid simulation methods to their research area.

O.-S. Kwon (✉)
Department of Civil and Mineral Engineering, University of Toronto, Toronto, ON, Canada
e-mail: os.kwon@utoronto.ca

V. Dertimanis
Institute of Structural Engineering, ETH Zürich, Zürich, Switzerland

M. Allen et al. (eds.), *Dynamic Substructures, Volume 4*, Conference Proceedings of the Society for Experimental Mechanics Series,
https://doi.org/10.1007/978-3-031-04094-8_10

10.2 Overview of Hybrid Simulation

10.2.1 The General Configuration

HS applies *substructuring* [7, 8] and establishes a *hybrid* model of a structure, by combining a numerical model, termed the *numerical substructure*, with a test specimen, termed the *experimental substructure*. The partitioning of the system into its constituent parts allows testing of the experimental substructure within the environment, in which it is designed to operate, rather than in an isolated fashion.

The underlying process is displayed in Fig. 10.1. Without loss of generality, it is assumed that the whole system is excited by a single input (force or displacement) and it can be split into one or more numerical substructures and one or more experimental substructures. At a given time instance, the numerical substructures are simulated under two input spaces: the former corresponds to the external input, whereas the latter predicted based on the response of the experimental substructures at the *common interface*. The structural response of the numerical substructures at the interface is then applied to the experimental substructures, by means of a *transfer system*, i.e., experimental equipment. The measured resisting forces, which are developed by the induced displacement of the experimental substructures, are then fed back as input to the numerical substructures for simulating the next time instance.

10.2.2 A Toy Example

Consider the paradigm of Fig. 10.2. A new tuned-mass damper (TMD) experimental prototype is available and the objective is to test its effectiveness on buildings, under diverse ground motions. Figure 10.2a displays the total structure, consisting of a simple, single-storey shear building with the TMD attached on the top. The equations of motion of the total structure in *relative coordinates* are

$$\begin{bmatrix} M & 0 \\ 0 & m \end{bmatrix} \begin{bmatrix} \ddot{u}_1(t) \\ \ddot{u}_2(t) \end{bmatrix} + \begin{bmatrix} K+k & -k \\ -k & k \end{bmatrix} \begin{bmatrix} u_1(t) \\ u_2(t) \end{bmatrix} = -\begin{bmatrix} M \\ m \end{bmatrix} \ddot{x}_g(t) \tag{10.1}$$

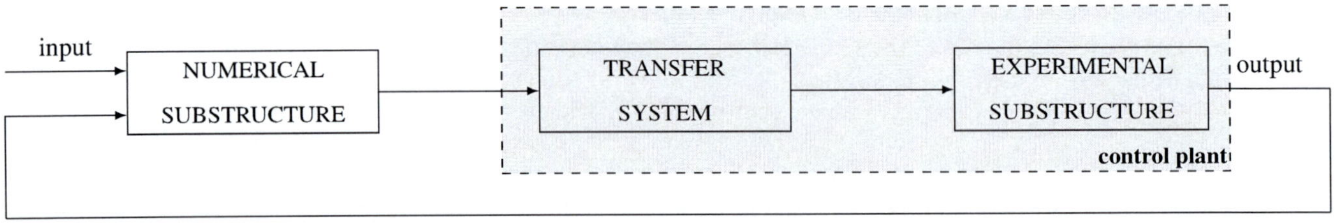

Fig. 10.1 Schematic layout of the HS framework

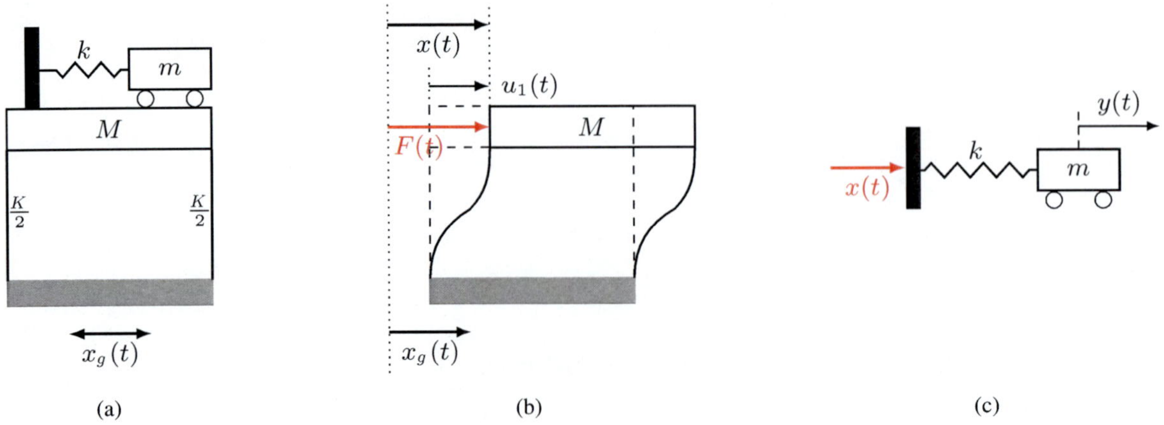

Fig. 10.2 A toy example: tuned-mass damper. (**a**) The total structure. (**b**) Numerical substructure. (**c**) Experimental substructure

where $x(t) = x_g(t) + u_1(t)$ and $y(t) = x_g(t) + u_2(t)$ are the corresponding *absolute* vibration displacement responses of M and m, respectively. The equation of motion for the numerical substructure can take the form

$$M\ddot{u}_1(t) + Ku_1(t) = -M\ddot{x}_g(t) + \underbrace{k[u_2(t) - u_1(t)]}_{F(t)} \tag{10.2}$$

where $F(t)$ represents the effects of the TMD on the shear frame, as shown in Fig. 10.2b. The equation of motion for the TMD reads

$$m\ddot{u}_2(t) + ku_2(t) - ku_1(t) = -m\ddot{x}_g(t) \tag{10.3}$$

or, shifting to absolute coordinates

$$m\ddot{y}(t) + ky(t) = kx(t) \tag{10.4}$$

Therefore, as indicated in Fig. 10.2c, the experimental substructure is excited by the absolute vibration displacement of the storey mass.

In conducting a HS for this toy example, Eq. (10.2) is used for simulating the numerical substructure. The excitation pertains to ground motion in terms of acceleration (typical in earthquake engineering applications), while the input force $F(t)$ is measured via an appropriate load cell attached to the transfer system (details below). The structural response $x(t)$ of the numerical substructure at the interface is then calculated (note: this configuration implies availability of $x_g(t)$) and subsequently applied to the TMD via the transfer system. It is noted that this is not the only feasible configuration for HS. Indicatively, one may conduct RTHS in the so-called *acceleration mode*, where the transfer system provides the motion in terms of acceleration $\ddot{x}(t)$. This relaxes the assumption of known $x_g(t)$. The force $F(t)$ can be alternatively measured by an accelerometer attached on the mass m, under the assumption that the latter is known. Refer to Dertimanis et al. [9, 10] for further details.

10.2.3 Distinction Between RTHS and PsDHS

The fundamental difference between RTHS and PsDHS is the time scale, at which the simulation is conducted. Conventional PsDHS is conducted at time scales that are usually two orders of magnitude slower that the actual time scale. This allows an inspection of the specimen during the HS loop, iterations to achieve force equilibrium within a time step, and iterations to impose predicted displacements when the experimental apparatus develops elastic deformation. Furthermore, highly nonlinear numerical model can be used as a numerical substructure as the analysis in each step does not need to be completed within a fixed time step. The disadvantage of this technique is that any load-rate sensitive features are not monitored, since the dynamics induced from inertial components are lost. Another disadvantage stems from the fact that displacement is applied over a period, during which the specimen is statically held until measurements are taken. Any effects developed during this period, such as stress relaxation, consequently, reduces accuracy.

In contrast to PsDHS, RTHS is executed in *hard real-time*. The HS loop is usually conducted in kHz rates and within this time all scheduled operations (i.e., numerical simulation, transfer system, and signal processing) must be successfully carried out, without losing any information. Such a setup is particularly suitable for structural elements that are rate-dependent such as acceleration and velocity. RTHS is generally applied to simulations where dynamics and vibration phenomena cannot be neglected. A carefully design RTHS provides significant insight on the actual behavior of a structure under study. However, as mentioned later in the description of the control plant, it poses significant challenges in the available hardware and software. A numerical substructure in RTHS is generally greatly simplified such that the numerical analysis can be completed within a specific time step.

10.2.4 Conditions to Apply HS

Hybrid simulation can be effectively used when the following two conditions are met: (1) there are bi-directional interactions between the numerically simulated substructures and the experimentally represented substructures, and (2) the experimentally substructures' response cannot be accurately numerically modeled. Regarding the first condition, in some problems, numerical substructures govern the boundary conditions of experimental substructures while the responses of the experimental substructure do not significantly influence the response of the numerical substructures. For example, a HS with one physical damper and tens of numerically modeled dampers in a building will not benefit from running the HS because the contributions of the physically tested damper to the global response are expected to be very limited. The second condition is rather obvious; if one can develop an accurate numerical model of a physical specimen, it is not necessary to run hybrid simulations. In some cases, though, modeling a structure with two or more numerical substructures with most suitable analysis tool, i.e., multi-platform hybrid simulation, can improve the accuracy of the simulation. To facilitate PsD or multi-platform hybrid simulations, the UT-SIM framework has been developed [11].

10.3 Indicative Case Studies

Three applications of HS are introduced in this section. In the first example, a real-time aeroelastic hybrid simulation is presented where the interaction between wind in wind tunnel and a model building is represented physically while the inertia, damping, and restoring forces are modeled numerically. In the second example, a steel column is modeled in a furnace while the rest of a building is modeled numerically to evaluate the response of a column subjected to a fire scenario. In the third example, four cast-steel yielding elements are modeled as physical substructures while the rest of a building is modeled numerically.

10.3.1 Real-Time Aeroelastic Hybrid Simulation

Dynamic interaction between fluid and deformable structure is very difficult to model numerically. While computational fluid dynamics (CFD) analysis can be used, it is not commonly used in the design process of high-rise buildings or long-span bridges due to the limited accuracy and high computational costs. In practice, a small scale model structure is built and tested in a wind tunnel to understand the wind force imposed to a structure through aerodynamic test or to understand vibration response of a structure through aeroelastic test. The aeroelastic test requires fabrication of a specimen, which can represent the dynamic properties, such as mass, stiffness, and damping, of lower mode vibration. Accurate development of such specimen requires many iterations, especially because it is difficult to physically model viscous damping in small scale.

To overcome the limitations of the current aeroelastic test methods, Kwon et al. [12] propose two designs of experimental equipment for real-time aeroelastic hybrid simulation (RTAHS) of a base-pivoting building model and a deck section model. Considering small force requirement, precision control, and high-frequency characteristics, it is proposed to adopt linear electric motors as an actuation system. Experimental validation is carried out for the base-pivoting building model [13]. Figure 10.3 shows the control loop for the RTAHS. More details on the RTAHS can be found in [13].

The main challenges in the RTAHS are the impact of delay in the linear motor's response. Because a structure is represented in small-scale in a wind tunnel, the natural frequency of the scaled model is relatively high approximately between 2 to 10 Hz. In such high-frequency response, the small delay in the motor's response leads to positive damping effect even after a delay compensation method was used. In Moni et al. [13], a mathematical expression is proposed, which relates actuator's delay and the apparent damping of the test system.

10.3.2 Real-Time Hybrid Fire Simulation of a Steel Column

Structural elements subjected to high temperature is also difficult to model numerically. For example, in a reinforced concrete element, the vapor pressure increases as external gas temperature increases. At some point, the vapor pressure leads to spalling of concrete cover. The material properties of concrete and steel drastically change as temperature increases.

Because of the challenges in the numerical modeling of structural elements subjected to high temperature, gas furnace is often used to evaluate structural performance in fire. In such tests, however, it is very difficult to model the structural system due to the limitations in the furnace. Running tests of one structural element in a furnace does not provide complete understanding on the behavior of the element as the conventional fire test cannot model the interaction between the tested specimen and the rest of the structure.

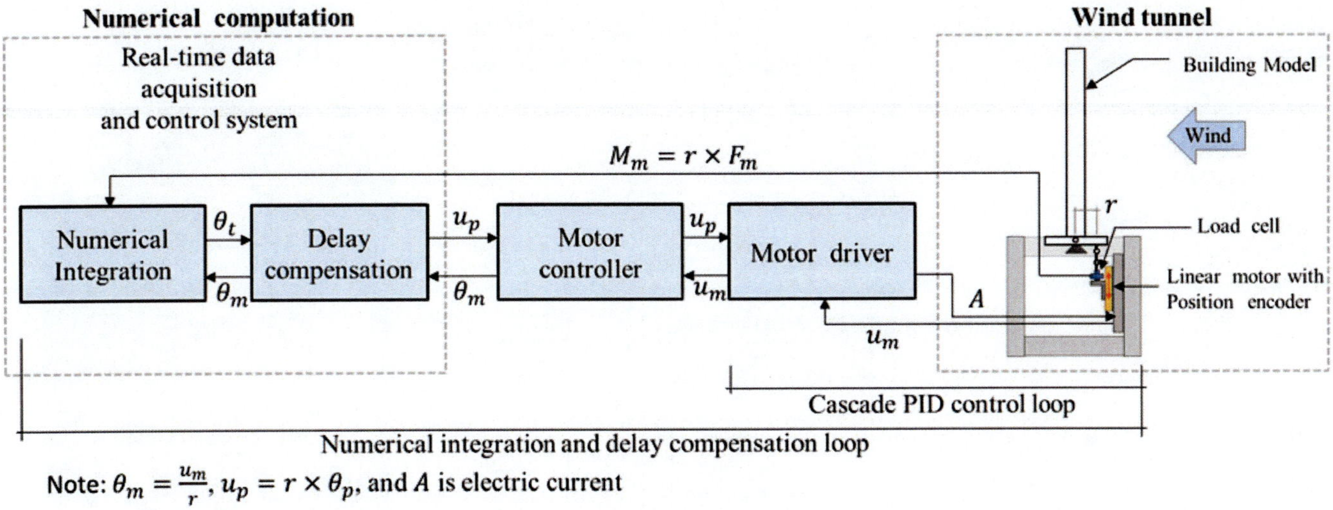

Fig. 10.3 Real-time aeroelastic hybrid simulation for a wind-excited building

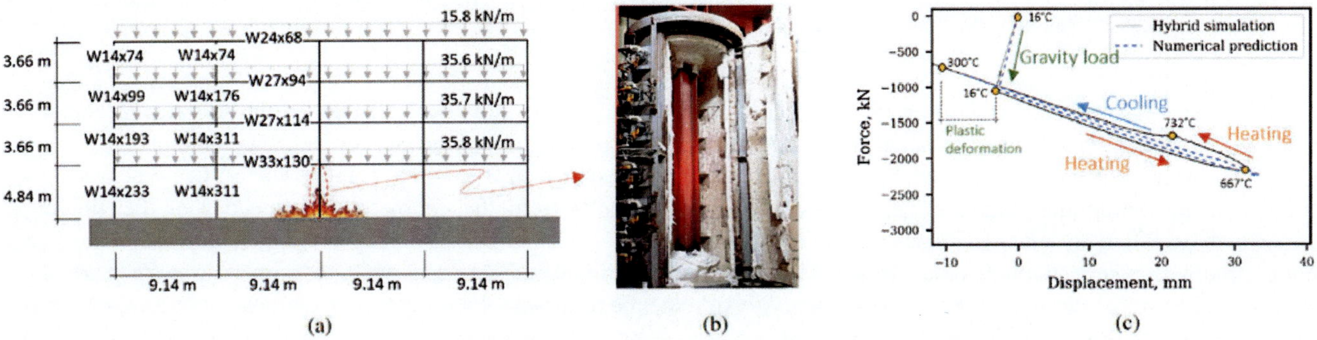

Fig. 10.4 Real-Time hybrid fire simulation of a steel column. (**a**) Building configuration. (**b**) Test specimen. (**c**) Specimen's response

To overcome the limitations in the conventional fire tests, Wang et al. [14, 15] propose real-time hybrid fire simulation method where a physical specimen's response in a furnace can be fed to a numerical model. A hybrid fire simulation method needs to be carried out in real-time because the gas temperature profile is defined as a function of time, and because the rate of heat conduction within a structural element cannot be scaled. By running the hybrid fire simulation method, it is possible to understand the interaction of a specimen subjected to fire with the rest of the structural system. The UT-SIM framework is used to run the hybrid simulation.

The main challenge in hybrid fire simulation is the accurate measurement of the specimen's deformation because the specimen is in a gas furnace. A specimen subjected to temperature load develops thermal expansion, which needs to properly controlled through a hydraulic actuator. Because the loading frame's stiffness is relatively small in comparison with the axial stiffness of a full-scale column, the elastic deformation of the loading is significant. Wang et al. [14] propose a method to infer the specimen's deformation based on the measured force and the loading frame's stiffness. Figure 10.4 illustrates the full hysteresis of a steel column's behavior obtained from a hybrid fire simulation. In a conventional fire test, a gravity load is held constant during the test. In reality, a specimen's thermal expansion deforms the rest of the structural system, which, in return, increases force demand to the specimen. Such interaction can be clearly captured through a hybrid fire simulation method.

10.3.3 Pseudo Dynamic Hybrid Simulation of a Steel Frame with Cast Steel Yielding Connectors

When a new structural element is developed to improve seismic performance of a building, the efficacy of the element is typically evaluated by running quasi-static cyclic tests, developing a numerical model of the element based on the test results,

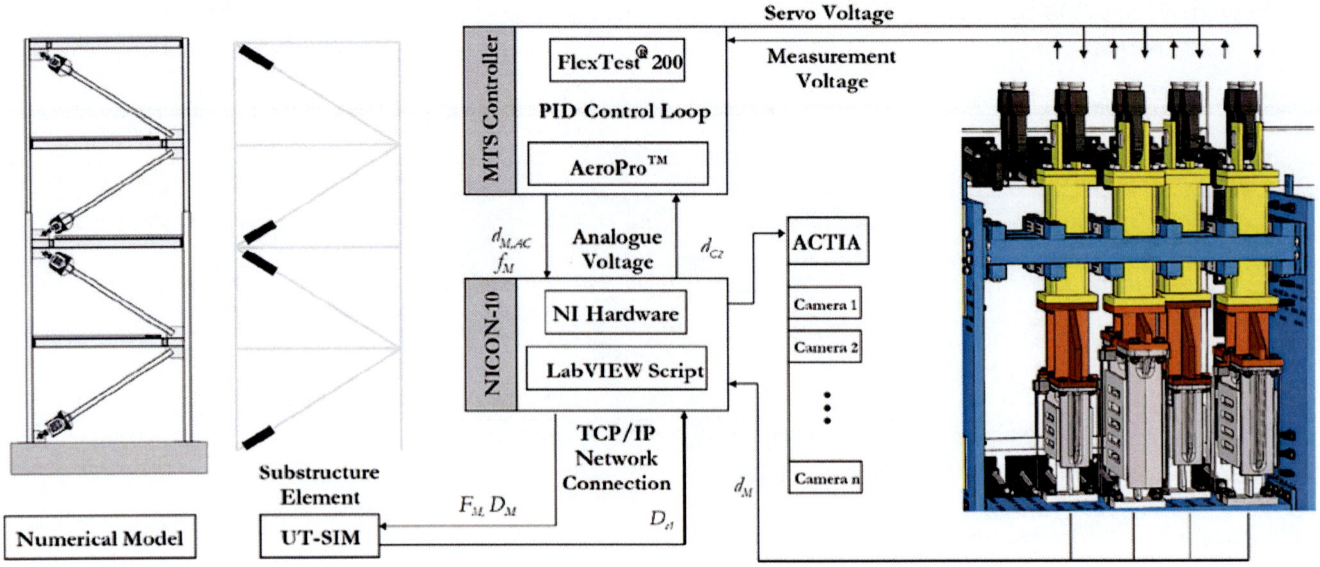

Fig. 10.5 PsDHS of a steel frame with cast steel yielding connectors

and running seismic simulations of the entire structural system including the numerical element of the structural elements. In many cases, though, development of an accurate numerical model of a highly nonlinear element is a daunting task.

One such example is the cast steel yielding connector (SYC) presented in Zhong et al. [16]. The SYC element shows elasto-plastic and hardening behavior. The hardening behavior results from the geometric nonlinearity of the yielding fingers when subjected to large deformation, and also hardening of the steel material. The SYC elements are connected to main structural system through bolts, which develops some frictional resistance as well. A numerical element of the complicated behavior is developed. Yet, it is necessary to validate the accuracy of the numerical elements by running hybrid simulations where multiple of SYCs were represented as physical specimens while the rest of the structural system is modeled numerically.

In Mortazavi et al. [17], for example, a four-storey braced frame is analyzed through PsDHS. In the simulation, four SYCs are modeled physically while the rest of the structural system is modeled numerically. All four SYCs are modeled physically to minimize the impact of numerical modeling error on the global response of the structure. UT10 equipment is used for the simulation, which can test up to ten uni-axially loaded elements simultaneously for a hybrid simulation. The UT-SIM framework is used as the main simulation framework. Figure 10.5 shows the overview of the test configuration. More details can be found in Mortazavi et al. [17].

10.4 The Control Plant

10.4.1 Main Components

A more detailed representation of the control plant is displayed in Fig. 10.6. It consists of an analog and a digital counterpart, which communicate via appropriate *digital-to-analog* and *analog-to-digital* converters (termed DAC and ADC, respectively). The digital component contains the *controller*; its purpose is to drive an *actuator* to a desired setpoint during each HS loop. This is accomplished by comparing a *reference signal* (the output of the numerical simulation) with the actuator's actual response, which is measured and fed back into the digital block. Observe that in Fig. 10.6a the feedback loop arrives from the output of the experimental substructure; this is because the actuator is firmly attached to it and thus the displacement of the actuator's piston is equal to the displacement of the substructure at the point of attachment.

The digital controller is typically provided by the vendor of the transfer system and it is characterized by a fixed *control law*. *Proportional–integral* (PI), or *proportional–integral–derivative* (PID) control are well-established techniques that are adequate for a wide range of transfer systems. However, the increased requirements of structural engineering tests (wider frequency response, response speed to the command signal, etc.) call for enhanced control measures. A popular such alternative to conventional PID is the *state variable control*, oftentimes referred to as *three variable control* (TVC). This

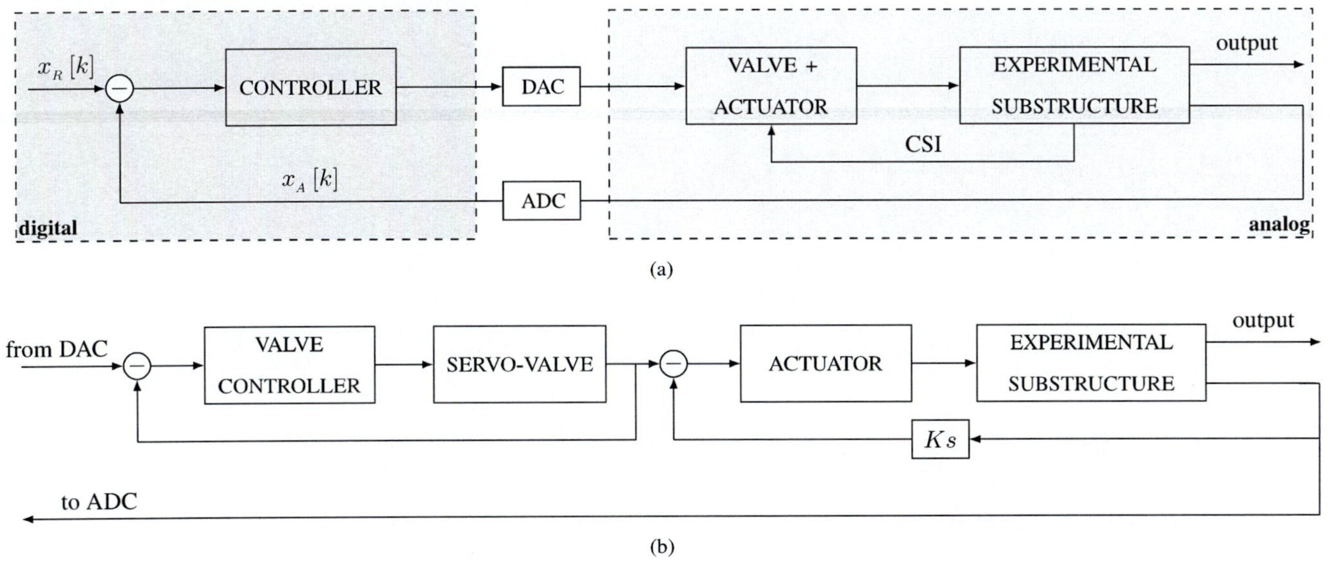

Fig. 10.6 The control plant. (**a**) General layout. The controller-structure interaction (CSI) loop generally applies only to RTHS. (**b**) The analog part of the control plant in more detail

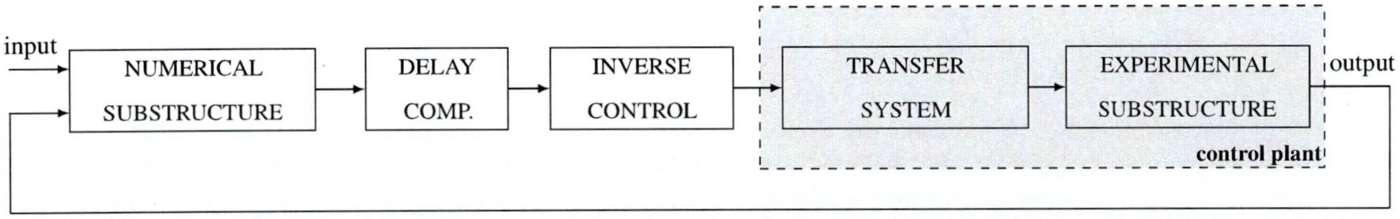

Fig. 10.7 The modified RTHS loop

technique takes into account all three variables of the motion (displacement, velocity, and acceleration), in a configuration that includes both feedback and feedforward branches. A detailed review of all the control alternatives for structural testing is provided by Plummer [18].

The analog component of the control plant includes the actuator and the experimental substructure. In structural testing, where usually the developed forces may reach the order of MN, actuators are hydraulic. The motion of a typical hydraulic actuator is determined by a hydraulic fluid that is supplied from a multi-stage servo-valve, which is also controlled, as shown in the left inner loop of Fig. 10.6b [18]. In the same block diagram, the right inner loop accounts for the *controller-structure interaction* (CSI), which is the result of the dynamic coupling (henceforth applies only to RTHS) between the actuator and the specimen. In Dyke et al. [19], it has been demonstrated that this coupling may be represented by a feedback loop that is proportional to velocity.

10.4.2 Two Problems

Focusing on RTHS, the accurate reproduction of the reference signal is challenging due to two main reasons, i.e., (1) the own dynamics of the transfer system, which act as a filter and modify the frequency content of the command; and (2) the *time delay*, which is inevitably introduced by the transfer system (valve, actuator, ADC, and DAC, controller), it is frequency-depended and causes a form of negative damping into the RTHS loop. The first issue is a typical control problem, while the second is a problem of prediction.

In tackling with these problems the RTHS loop of Fig. 10.1 can be modified as shown in Fig. 10.7. Two additional blocks have been now added, which account for the compensation of the transfer delay and the transfer system dynamics, respectively. The former accepts a reference signal from the numerical substructure, $x_R[k]$ and returns a prediction at Δ time instants forward, i.e., $x_R[k+\Delta]$. This may be accomplished by polynomial extrapolation [5, 20, 21] via the formula

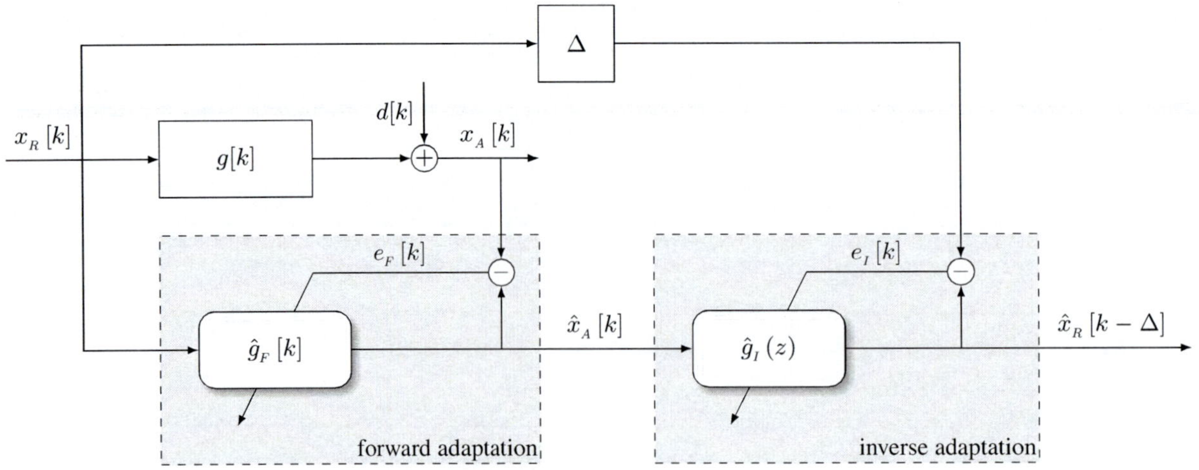

Fig. 10.8 Data-driven estimation of the inverse controller. From [22]

$$x_R[k+\Delta] = \sum_{i=0}^{n_P} p_i x_R[k - i\Delta] \tag{10.5}$$

where n_P is the polynomial order and p_i the polynomial coefficients, calculated via the Lagrange basis function and are predefined for a given order.

The role of the inverse controller is to cancel the dynamics of the control plant in a feedforward manner. If the latter is characterized by an equivalent digital transfer function $G(z)$, then the inverse controller should be described by the inverse transfer function, $G^{-1}(z)$, forcing the cascade to perform as a Δ-delayed pure impulse.

10.4.3 Data-Driven RTHS

The proper estimation of the inverse controller is not an easy task. One method is described in Simpson et al. [22] and it has the advantage of being purely *data-driven*, i.e., it does not rely on any prior information for the transfer system; it uses only the reference and the achieved signals. The general framework is displayed in Fig. 10.8 and it consists of two stages. In the first stage, an finite impulse response (FIR) estimate of $g[k]$, $\hat{g}_F[k]$, is obtained, where $g[k]$ is the inverse $\mathcal{Z}$ transform of the control plant's digital transfer function. Then, in the second stage, an FIR estimate of the inverse of $g[k]$, $\hat{g}_I[k]$, is obtained.

The adaptation process supplies a reference signal of favorable properties [22, 23] to the control plant. This signal, together with the noise-corrupted achieved response of the plant, $x_A[k]$ are used by the forward adaptation block, for estimating $g[k]$. When the forward and inverse adaptation are carried out in *synchronous mode*, the instantaneous estimate of the achieved response, $\hat{x}_A[k]$, together with a *delayed* version of the original reference signal, are fed to the inverse adaptation block and $\hat{g}_I[k]$ is tuned for describing the achieved-(delayed) reference signal dynamics. The output of this block is an estimate of $x_R[k-\Delta]$. Upon successful execution of the two stages, the inverse FIR filter $\hat{g}_I[k]$ is copied before the control plant, as shown in Fig. 10.7. It is noted that more sophisticated algorithms can be implemented as adaptive inverse controllers, as the modified filtered-X one presented in Dertimanis et al. [9].

10.5 Conclusions

This paper presents a high-level overview on RTHS and PsDHS. In hybrid simulations, numerical substructures are integrated with experimental substructures such that the physical substructures behavior can be used in the simulation of a structural system. The simulation methods can overcome the limitations of fully numerical simulation or fully experimental tests on physical specimens. RTHS and PsDHS have their different application areas; the former is suitable for testing rate-dependent elements and relatively simplified numerical substructures, and the latter is suitable for rate-independent elements and can accommodate highly nonlinear and realistic numerical substructures. A few case studies are introduced to present the potential of the methods in various loading scenarios.

References

1. Song, W. , Chang, C.-M., Dertimanis, V.K.: Editorial: recent advances and applications of hybrid simulation. Front. Built Environ. **6**, 203 (2020)
2. Pegon, P.: Continuous PsD testing with substructuring. In: Bursi, O.S., Wagg, D. (eds.) Modern Testing Techniques for Structural Systems, chap. 5, pp. 197–257. Springer, Vienna (2008)
3. Bayer, V. , Dorka, U.E. , Füllekrug, U., Gschwilm, J.: On real-time pseudo-dynamic sub-structure testing: algorithm, numerical and experimental results. Aerospace Sci. Technol. **9**(3), 223–232 (2005)
4. Nakashima, M., Kato, H., Takaoka, E.: Development of real time pseudodynamic testing. Earthq. Eng. Struct. Dyn. **21**, 79–92 (1992)
5. Wallace, M., Wagg, D., Neild, S.: An adaptive polynomial based forward prediction algorithm for multi-actuator real-time dynamic substructuring. Proc. R Soc. A Math. Phys. Eng. Sci. **461**(2064), 3807–3826 (2005)
6. Carrion, J.E., B. F. Spencer, J.: Real-time hybrid testing using model-based delay compensation. In: 4th International Conference on Earthquake Engineering, Taipei, p. 299 (2006)
7. de Klerk, D., Rixen, D.J., Voormeeren, S.N.: General framework for dynamic substructuring: history, review and classification of techniques. AIAA J. **46**(5), 1169–1181 (2008)
8. Allen, M.S., Rixen, D., van der Seijs, M., Tiso, P., Abrahamsson, T., Mayes, R.L.: Substructuring in Engineering Dynamics: Emerging Numerical and Experimental Techniques. Springer, Cham (2020)
9. Dertimanis, V., Mouzakis, H., Psycharis, I.: On the acceleration-based adaptive inverse control of shaking tables. Earthq. Eng. Struct. Dyn. **44**(9), 1329–1350 (2015)
10. Dertimanis, V. , Mouzakis, H., Psycharis, I.: On the control of shaking tables in acceleration mode: an adaptive signal processing framework. In: Taucer, F., Apostolska, R. (eds.) Experimental Research in Earthquake Engineering: EU-SERIES Concluding Workshop, pp. 159–172. Springer, Cham (2015)
11. Huang, X., Kwon, O.-S.: A generalized numerical/experimental distributed simulation framework. J. Earthquake Eng. **24**(4), 682–703 (2020)
12. Kwon, O.-S. , Kim, H.-K. , Jeong, U.Y., Hwang, Y.-C., Moni, M.: Design of Experimental Apparatus for Real-Time Wind-Tunnel Hybrid Simulation of Bridge Decks and Buildings, pp. 235–245. ASCE, Baltimore (2019)
13. Moni, M., Hwang, Y., Kwon, O.-S., Kim, H.-K., Jeong, U.Y.: Real-time aeroelastic hybrid simulation of a base-pivoting building model in a wind tunnel. Front. Built Environ. **6**, 157 (2020)
14. Wang, X., Kim, R.E., Kwon, O.-S., Yeo, I.: Hybrid simulation method for a structure subjected to fire and its application to a steel frame. J. Struct. Eng. **144**(8), 04018118 (2018)
15. Wang, X., Kim, R.E., Kwon, O.-S., Yeo, I.-H., Ahn, J.-K.: Continuous real-time hybrid simulation method for structures subject to fire. J. Struct. Eng. **145**(12), 04019152 (2019)
16. Zhong, C., Binder, J., Kwon, O.-S., Christopoulos, C.: Full-scale experimental testing and postfracture simulations of cast-steel yielding connectors. J. Struct. Eng. **146**(12), 04020261 (2020)
17. Mortazavi, P., Kwon, O.-S., Christopoulos, C.: Four-element pseudodynamic hybrid simulation of a steel frame with cast steel yielding connectors under earthquake excitations. J. Struct. Eng. **148**, 04021255 (2021)
18. Plummer, A.R.: Control techniques for structural testing: a review. Proc. Inst. Mech. Eng. I J. Syst. Control Eng. **221**(2), 139–169 (2007)
19. Dyke, S.J., Spencer, B.F., Quast, P., Sain, M.K.: Role of control-structure interaction in protective system design. J. Eng. Mech. **121**(2), 322–338 (1995)
20. Horiuchi, T., Inoue, M., Konno, T., Namita, Y.: Real-time hybrid experimental system with actuator delay compensation and its application to a piping system with energy absorber. Earthq. Eng. Struct. Dyn. **28**(10), 1121–1141 (1999)
21. Darby, A.P., Blakeborough, A., Williams, M.S.: Improved control algorithm for real-time substructure testing. Earthq. Eng. Struct. Dyn. **30**(3), 446–448 (2001)
22. Simpson, T., Dertimanis, V.K., Chatzi, E.N.: Towards data-driven real-time hybrid simulation: adaptive modeling of control plants. Front. Built Environ. **6**, 172–187 (2020)
23. Dertimanis, V., Mouzakis, H., Psycharis, I.: Real-time shaking table adaptive modeling under acceleration control: effects of training signals and specimen dynamics. In: Proceedings of the Sixth World Conference on Structural Control and Monitoring, Barcelona, pp. 996–1008 (2014)

Chapter 11
Identification of Bolted Joint Properties Through Substructure Decoupling

Jacopo Brunetti, Walter D'Ambrogio, Matteo Di Manno, Annalisa Fregolent, and Francesco Latini

Abstract Substructure decoupling techniques, defined in the frame of Frequency Based Substructuring, allow to identify the dynamic behaviour of a structural subsystem starting from the known dynamics of the coupled system and from information about the remaining components. The problem of joint identification can be approached in the substructuring framework by decoupling jointed substructures from the assembled system. In this case, information about the coupling DoFs of the assembled structure is necessary and this could be a problem if the interface is inaccessible for measurements. Expansion techniques can be used to obtain the dynamics on inaccessible (interface) DoFs starting from accessible (internal) DoFs. A promising technique is the System Equivalent Model Mixing (SEMM) that combines numerical and experimental models of the same component to obtain a hybrid model. This technique has been already used in an iterative coupling–decoupling procedure to identify the linear dynamic behaviour of a joint, with a Virtual Point description of the interface. In this work, a similar identification procedure is applied to the Brake Reus Beam benchmark to identify the linear dynamic behaviour of a three bolted connection at low levels of excitation. The joint is considered as a third independent substructure that accounts for the mass and stiffness properties of the three bolts, thus avoiding singularity in the decoupling process. Instead of using the Virtual Point Transformation, the interface is modelled by performing a modal condensation on remote points allowing deformation of the connecting surfaces between subcomponents. The purpose of the study is to highlight numerical and ill-conditioning problems that may arise in this kind of identification.

Keywords Experimental substructuring · Decoupling · Joint identification · Model mixing · Deformable connecting surfaces

11.1 Introduction

Most of the engineering structures are composed of different parts connected together using bolted joints, tape and fir-tree joints. These joints can modify locally the mass, stiffness and damping properties of the assembled structure [1]. For this reason, it is useful to identify the characteristics of the joint in order to understand its effects on the dynamics of the whole system. This could allow to use the identified joint in other analyses to evaluate the dynamics of assemblies that present the same type of connection. Substructuring techniques are useful for both the identification and for the use of the identified joint in other configurations. A general framework on dynamic substructuring is provided by De Klerk et al. in [2], where the approaches in physical, modal and frequency domain are presented supposing that the systems can be described by linear models. Substructuring techniques are broadly used because, among other advantages, they allow to combine experimental and numerical models [3, 4]. Dealing with experimental data, the substructuring approach in frequency domain, called Frequency Based Substructuring (FBS) [5] is the most advantageous. The substructuring approach allows to solve both the direct and the inverse problem, i.e. coupling and decoupling, respectively. The substructure coupling [6] consists in the identification of the dynamic behaviour of an assembled system starting from the known dynamic behaviour of the component subsystems. The DoFs of the assembled system can be partitioned into internal DoFs (not belonging to the couplings) and

J. Brunetti · W. D'Ambrogio
Dipartimento di Ingegneria Industriale e dell'Informazione e di Economia, Università dell'Aquila, L'Aquila, Italy
e-mail: jacopo.brunetti@univaq.it; walter.dambrogio@univaq.it

M. D. Manno (✉) · A. Fregolent · F. Latini
Dipartimento di Ingegneria Meccanica e Aerospaziale, Università di Roma La Sapienza, Rome, Italy
e-mail: matteo.dimanno@uniroma1.it; annalisa.fregolent@uniroma1.it; francesco.latini@uniroma1.it

M. Allen et al. (eds.), *Dynamic Substructures, Volume 4*, Conference Proceedings of the Society for Experimental Mechanics Series,
https://doi.org/10.1007/978-3-031-04094-8_11

coupling DoFs. Inversely, substructure decoupling [7–11] consists in the identification of an unknown component subsystem from the known dynamic behaviour of the assembled system and of the residual part of the assembly.

Substructure decoupling can be used to perform the identification of the characteristics of a joint considered as a subsystem connected to the other physical subsystems of the assembly [12–14]. In order to obtain the dynamic behaviour of the joint substructure at coupling DoFs, substructure decoupling requires the information of the assembled structure at least, on the coupling DoFs. This could be a problem if the joint interface is not accessible for measurements, but it can be overcome by using for example System Equivalent Model Mixing (SEMM) [4]. It performs coupling and decoupling operations between the numerical and the experimental model of the same component to obtain a hybrid model in which the experimental dynamics measured at accessible DoFs is expanded on inaccessible DoFs. In [15], the inaccessible dove-tail joint of a bladed disk structure is analysed and the SEMM technique is implemented as an iterative procedure [16] to identify the properties of the connection. In that application, even if there is not a physical joint, the connection itself is considered as a third independent substructure and the dynamics at the connecting DoFs is obtained by using the Virtual Point Transformation (VPT) [17].

In this work, a modification to the joint identification procedure described previously is presented. The SEMM models of each substructure and of the coupled assembly are derived as usual. Nevertheless, the following modifications are introduced with reference to a bolted joint. First, in the iterative procedure, the initial guess for the joint is modelled as a lumped parameter system composed of two masses connected through elastic elements. In particular, the mass property is not fictitious, but it is derived from the mass of the bolts in the real junction. Secondly, in order to account for the deformation of the connection interface, remote points are used to represent its dynamics instead of using the virtual points. This is done to deal with interface regions that are of finite length with respect to the assembled structure and cannot be considered as rigid. The devised procedure is applied to identify the properties of the bolted joint of a numerically simulated Brake Reuss Beam. This kind of structure was designed and used by Brake [1] to study friction in bolted joints, and it is composed of two identical beams with square cross-section that are jointed together through three bolts at their common end. The purpose of the study is to highlight numerical and ill-conditioning problems that may arise in this kind of identification.

11.2 Theoretical Background

In this section, the methods used to obtain the Frequency Response Functions (FRFs) of the models are described. In particular, the theory of LM-FBS and the SEMM method are presented.

11.2.1 Coupling Using LM-FBS

Considering a number n of uncoupled substructures, the equation of motion that describes the r-th substructure in the frequency domain, can be written as

$$\{u(\omega)\}^{(r)} = [Y(\omega)]^{(r)} \left(\{f(\omega)\}^{(r)} + \{g(\omega)\}^{(r)}\right) \tag{11.1}$$

where $\{u(\omega)\}^{(r)}$ is the response vector, $\{f(\omega)\}^{(r)}$ is the external force vector, $\{g(\omega)\}^{(r)}$ is the vector of connecting forces with other subsystems and $[Y(\omega)]^{(r)}$ is the FRF.

The equations of motion of the n subsystems to be coupled can be written in diagonal form as

$$\{u\} = [Y] \left(\{f\} + \{g\}\right) \tag{11.2}$$

with

$$[Y] = \begin{bmatrix} [Y]^{(1)} & & \\ & \ddots & \\ & & [Y]^{(n)} \end{bmatrix}, \quad \{u\} = \begin{Bmatrix} \{u\}^{(1)} \\ \vdots \\ \{u\}^{(n)} \end{Bmatrix}, \quad \{f\} = \begin{Bmatrix} \{f\}^{(1)} \\ \vdots \\ \{f\}^{(n)} \end{Bmatrix}, \quad \{g\} = \begin{Bmatrix} \{g\}^{(1)} \\ \vdots \\ \{g\}^{(n)} \end{Bmatrix} \tag{11.3}$$

where the explicit frequency dependence is omitted.

Substructures can be coupled by imposing the compatibility and equilibrium conditions between the connecting DoFs. The compatibility condition at the connecting DoFs implies that any pair of matching DoFs $u_l^{(r)}$ and $u_m^{(s)}$ i.e. DoF l on subsystem r and DoF m on subsystem s have the same displacement, that is $u_l^{(r)} - u_m^{(s)} = 0$. This condition can be generally expressed by introducing the signed Boolean matrix $[B]$:

$$[B]\{u\} = \{0\} \tag{11.4}$$

The equilibrium condition states that for the same pairs of connecting DoFs, the interface forces must be equal and opposite in sign, i.e. $g_l^{(r)} + g_m^{(s)} = 0$. In LM-FBS formulation, a set of unknown Lagrange multipliers $\{\lambda\}$ is introduced such that the equilibrium conditions can be rewritten using again the Boolean matrix $[B]$:

$$[B]^T\{\lambda\} = -\{g\} \tag{11.5}$$

By substituting the interface forces $\{g\}$ into Eq. (11.5), the following system of equations is obtained:

$$\begin{cases} \{u\} = [Y](\{f\} - [B]^T\{\lambda\}) \\ [B]\{u\} = \{0\} \end{cases} \tag{11.6}$$

It is possible to eliminate $\{\lambda\}$ and obtain the single line equation of LM-FBS:

$$\{u\} = \left[\bar{Y}\right]\{f\} \tag{11.7}$$

in which $[\bar{Y}]$ is the dually-assembled frequency response function for the system:

$$\left[\bar{Y}\right] = [Y] - [Y][B]^T([B][Y][B]^T)^{-1}[B][Y] \tag{11.8}$$

11.2.2 Decoupling Using LM-FBS

Substructure decoupling allows to obtain the dynamic behaviour of an unknown subsystem A starting from the known behaviour of the assembled system AB and from information about the remaining component B, also referred as residual substructure. This procedure can be useful, for example, if substructure A is not accessible for direct measurements. In LM-FBS formulation, the decoupling process can be implemented by adding to the assembled system AB a fictitious substructure with an FRF opposite in sign to that of the residual subsystem B. This can be achieved by setting the following decoupled FRF matrix in Eq. (11.8):

$$[Y] = \begin{bmatrix} [Y]^{AB} & \\ & -[Y]^B \end{bmatrix} \tag{11.9}$$

When dealing with coupling operations, the compatibility and equilibrium conditions must be imposed on the coupling (or connecting) DoFs that are shared between substructures. This requires that both responses and forces must be determined on each coupling DoF thus resulting in square interface matrices $[B][Y][B]^T$.

Nevertheless, in decoupling operations the known structure AB and substructure B share both coupling and internal DoFs, thus equilibrium and compatibility conditions can be imposed on internal and connecting DoFs resulting in an overdetermined problem in which the interface is addressed as extended. Furthermore, compatibility and equilibrium can be specified on different DoFs, thus resulting in distinct Boolean matrices $[B_C]$ and $[B_E]$ for displacement and equilibrium. For this reason, Eq. (11.8) needs to be adapted by defining the coupled FRF matrix $[\bar{Y}]$ in which the pseudo-inverse operator $(\bullet)^+$ is used for the inversion of the interface matrix $[B_C][Y][B_E]^T$.

$$\left[\bar{Y}\right] = [Y] - [Y][B_E]^T([B_C][Y][B_E]^T)^+[B_C][Y] \tag{11.10}$$

11.2.3 System Equivalent Model Mixing

The SEMM is a technique based on the LM-FBS formulation that expands the information measured on some DoFs of the experimental model on some DoFs of the numerical one by performing coupling and decoupling operations. In this way, the considered component is described by a hybrid model in which it is possible to have information on inaccessible DoFs by exploiting measured data of the experimental model. In order to briefly explain how this method is derived, the DoFs of the component can be classified in different sets [15]:

- $\{u_c\}$: Compatibility DoFs where response is measured.
- $\{u_e\}$: Equilibrium DoFs where impacts are applied.
- $\{u_o\}$: Other DoFs that are measured but are only used for validation purposes.
- $\{u_b\}$: Boundary (or coupling) DoFs that are assumed to be inaccessible for measurements and excitation.

The different models involved in the expansion process are described below:

- *Parent model* $[Y]^{par}$: Numerical model defined on the global set of DoFs $\{u_g\}$:

$$[Y]^{par} = \begin{bmatrix} [Y_{cc}] & [Y_{ce}] & [Y_{co}] & [Y_{cb}] \\ [Y_{ec}] & [Y_{ee}] & [Y_{eo}] & [Y_{eb}] \\ [Y_{oc}] & [Y_{oe}] & [Y_{oo}] & [Y_{ob}] \\ [Y_{bc}] & [Y_{be}] & [Y_{bo}] & [Y_{bb}] \end{bmatrix}^{par} \quad \text{and} \quad \{u_g\} = \begin{Bmatrix} \{u_c\} \\ \{u_e\} \\ \{u_o\} \\ \{u_b\} \end{Bmatrix} \tag{11.11}$$

- *Overlay model* $[Y]^{ov}$: Experimental model of the component obtained by measuring on the compatibility DoFs $\{u_c\}$ when exciting at the equilibrium DoFs $\{u_e\}$.

$$[Y]^{ov} = [Y_{ce}]^{ov} \tag{11.12}$$

- *Removed model* $[Y]^{rem}$: It is a numerical condensed form of the parent model used to remove the parent dynamics from the component. In the so-called Extended SEMM presented in [4], the removed model is defined on the global set of DoFs and coincides with the parent model:

$$[Y]^{rem} = \left[Y_{gg}\right]^{par} \tag{11.13}$$

- *Hybrid model* $[Y]^{SEMM}$: Result of SEMM expansion. It is defined on the same DoFs of the parent model. For the Extended SEMM equation, specified for the global set of DoFs $\{u_g\}$, it is [15]:

$$\left[Y\right]^{SEMM} = \left[Y_{gg}\right]^{par} - \left[Y_{gg}\right]^{par} \left(\left[Y_{cg}\right]^{par}\right)^{+}([Y_{ce}]^{par} - [Y_{ce}]^{ov})(\left[Y_{ge}\right]^{par})^{+} \left[Y_{gg}\right]^{par} \tag{11.14}$$

The expansion technique is described in Fig. 11.1.

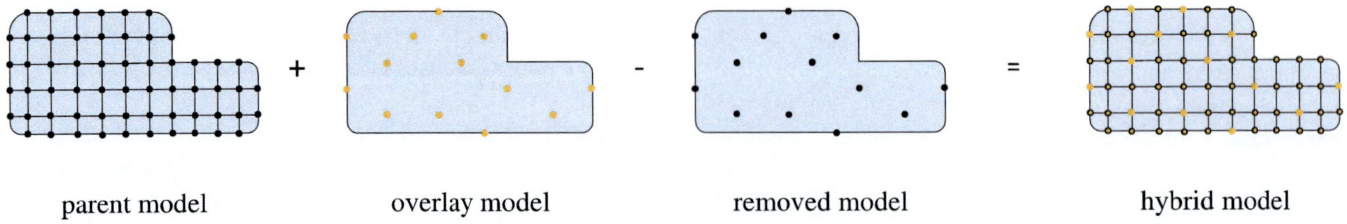

Fig. 11.1 SEMM expansion technique. The dynamics measured on the DoFs of the experimental overlay model, are overlapped in the numerical model that provides the DoF set of the component. The dynamics of the numerical model are then subtracted by decoupling the removed model

11.3 Joint Identification Procedure

In this section, the identification procedure of the joint connecting substructure A and substructure B is described considering that the connecting interface is not accessible for measurements. The joint is considered as a third independent substructure J with given mass and stiffness properties. Thus, its behaviour can be obtained by decoupling the dynamics of substructures A and B from the dynamic behaviour of the assembled structure AB as shown in Fig. 11.2. As seen from Fig. 11.2, substructures A and B have both internal and boundary DoFs, and the joint J is defined over the same boundary DoFs of A and B, which are not known if the interface is not accessible for measurements. Thus, in order to identify the joint, the dynamic behaviour of the assembled structure must be known at the boundary DoFs $\{u_b\}^A$ and $\{u_b\}^B$. For this reason, it is necessary to obtain the SEMM models either of the assembled structure and also of the decoupled substructures A and B. The FRF of the joint at iteration i $[Y]_i^J$ is used together with the SEMM models of substructures A and B to obtain the decoupled FRF matrix $[Y]_{i+1}$:

$$[Y]_{i+1} = \begin{bmatrix} [Y]^{A,SEMM} & & \\ & [Y]_i^J & \\ & & [Y]^{B,SEMM} \end{bmatrix} \tag{11.15}$$

used in Eq. (11.8) to obtain the parent model of the assembled structure $[Y]_{i+1}^{AB,par}$ that contains the boundary DoFs but lacks of the real dynamics introduced by the joint. This information can be introduced in the model $[Y]_{i+1}^{AB,par}$ by performing a SEMM expansion using the experimental Overlay model $[Y]^{AB,ov}$ thus obtaining the Hybrid model $[Y]_{i+1}^{SEMM,par}$. The FRF of the joint substructure at iteration $i+1$ is then obtained by decoupling the SEMM models of substructures A and B from the assembly AB:

$$[Y]_{i+1}^J = \begin{bmatrix} [Y]_{i+1}^{AB,SEMM} & & \\ & -[Y]^{A,SEMM} & \\ & & -[Y]^{B,SEMM} \end{bmatrix} \tag{11.16}$$

Note that in order to start the procedure, an initial guess for the joint is estimated by assigning mass and stiffness matrices M_J and K_J and its FRF $[Y]_0^J$. The iterative algorithm stops when the following convergence criterion is satisfied:

$$\frac{\left|[Y_{ce}]_{i+1}^{AB,par} - [Y_{ce}]^{AB,ov}\right|_2}{\left|[Y_{ce}]^{AB,ov}\right|_2} < \varepsilon \tag{11.17}$$

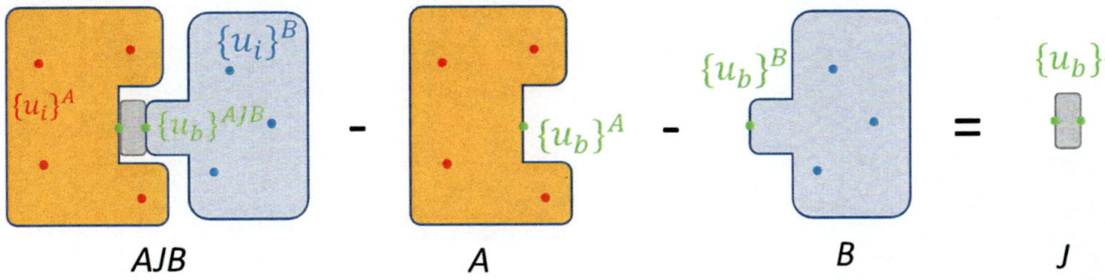

Fig. 11.2 Joint identification through substructure decoupling. The internal DoFs are indicated as $\{u_i\}$; $\{u_b\}$ are the boundary DoFs

11.4 Brake Reuss Beam Model

The joint identification procedure is applied to a Brake Reuss Beam: it is composed of two identical beams with square cross-section that have half-height extension near one of their ends. The two beams can be overlapped in correspondence of the half-height extensions in order to be jointed with three bolts, thus forming a lap joint as shown in Fig. 11.3. As seen from Fig. 11.3, substructure A is clamped at one end both in the coupled and uncoupled configuration, while substructure B is always unconstrained. The beams are made of steel alloy AISI-304 and the dimensions of the assembly are 720.9 x 25 x 25 mm. Each bolt consists of a screw M8x40 hexagon head and a M8 nut with washer under the bolt head and the nut. Both the beams and the assembly have been modelled in Ansys in order to build the parent and the overlay models to be used in the identification procedure outlined in Sect. 11.3. In the present work, the overlay model is obtained from numerical simulations instead of experimental tests in order to validate the identification procedure. Some important aspects need to be pointed out regarding the models of the beams and of the assembly.

11.4.1 SEMM Models

A linear finite element model of each substructure is developed in Ansys. The interface surface of each beam is divided into five areas and the nodes belonging to each area are connected to a dependent remote point located in the middle of the area using Multi-Point Constraint equations [18, 19] as shown in Fig. 11.4. This assumption allows to retain the local flexibility of the interface.

Since the SEMM model of each beam is needed, the following sets of DoFs are chosen:

$$\{u_g\}^A = \begin{Bmatrix} \{u_c\}^A \\ \{u_e\}^A \\ \{u_o\}^A \\ \{u_b\}^A \end{Bmatrix}, \quad \{u_g\}^B = \begin{Bmatrix} \{u_c\}^B \\ \{u_e\}^B \\ \{u_o\}^B \\ \{u_b\}^B \end{Bmatrix} \tag{11.18}$$

Note that the boundary DoFs $\{u_b\}^A$ and $\{u_b\}^B$ are located on the connecting surfaces and they are not accessible for measurement and excitation in the assembled configuration.

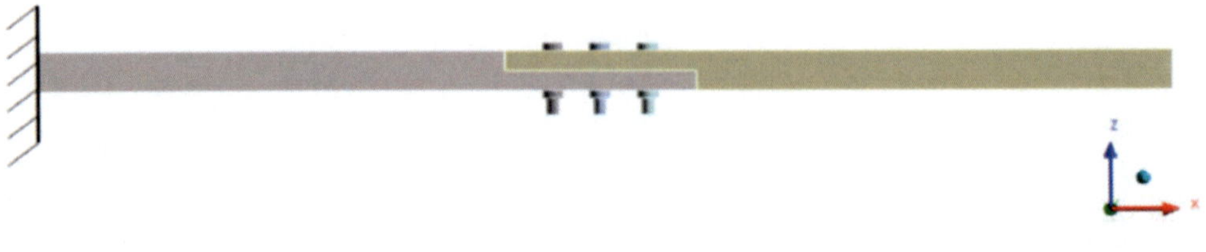

Fig. 11.3 Brake reuss beam

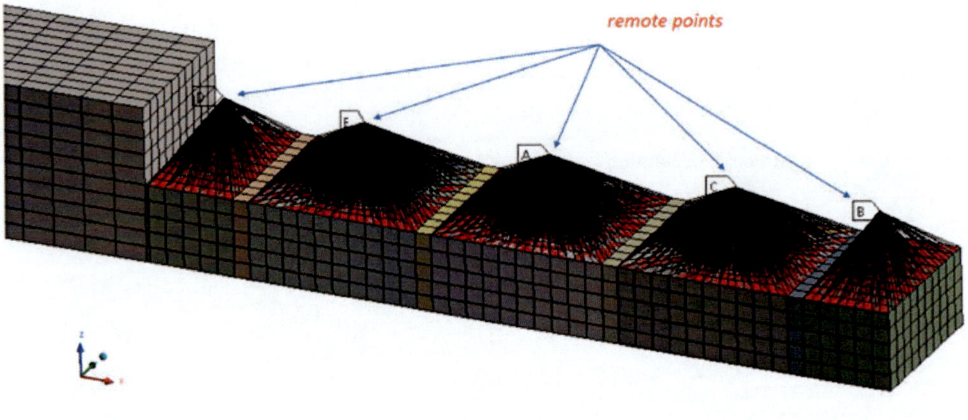

Fig. 11.4 Remote points used to account for connecting surface deformation

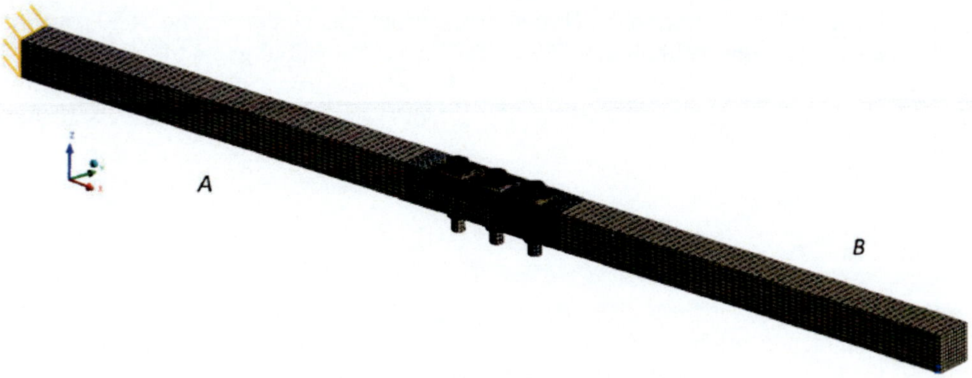

Fig. 11.5 FE model of the brake reus beam. Beam A is clamped at its end

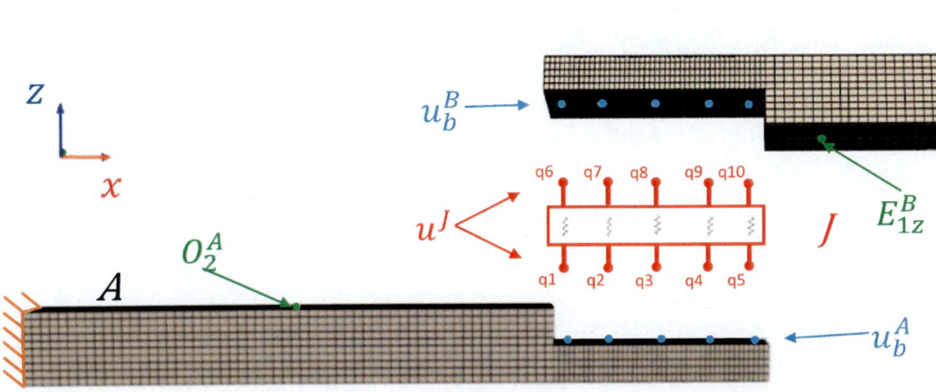

Fig. 11.6 Generation of $[Y]^{AJB,par}$. The SEMM models of substructures A and B are coupled with the joint. The joint substructure is defined on the translational and rotational DoFs of the points $q1$–$q10$

Regarding the SEMM model of the Brake Reuss Beam, in order to simulate the experimental Overlay Model of the assembled structure $[Y]^{AB,ov}$, a numerical FEM model is first generated (Fig. 11.5). The remote points, previously defined on the connecting surfaces of the beams, are deleted and the FE models of the three bolts are introduced. The built-in Bolt Pretension Tool of Ansys is used to apply an axial load to each bolt in order to simulate a tightening torque T of 14 Nm. The modes of the Brake Reuss Beam are computed with a nonlinear pre-stressed analysis that is able to account for bolt preload. The Mode Superposition Method implemented in [20] is used to numerically generate the overlay model of the assembled structure, which is defined on the same measurement DoFs $\{u_c\}^{AB}$ and excitation DoFs $\{u_e\}^{AB}$ of the two beams, i.e.:

$$\{u_c\}^{AB} = \begin{Bmatrix} \{u_c\}^A \\ \{u_c\}^B \end{Bmatrix} \qquad \{u_c\}^{AB} = \begin{Bmatrix} \{u_e\}^A \\ \{u_e\}^B \end{Bmatrix} \tag{11.19}$$

Regarding the parent model $[Y]^{AJB,par}$ shown in Sect. 11.3, it is generated as described in Fig. 11.6 at each iteration: The SEMM models of the beams A and B are coupled with the joint substructure J by imposing compatibility and equilibrium conditions between the boundary DoFs of the two beams (translational and rotational DoFs of the blue dots) and the coupling DoFs of the joint (translational and rotational DoFs of the red dots). Thus, the DoFs of the parent model can be collected as follows:

$$\{u_g\}^{AB} = \begin{Bmatrix} \{u_g\}^A \\ \\ \{u_g\}^B \end{Bmatrix} \tag{11.20}$$

Note that, in Eq. (11.20) the DoFs of the joint $\{u\}^J$ are not reported, as they are coincident with the boundary DoFs of the beams.

The joint is defined on 60 DoFs (both translational and rotational), which can be considered as 30 pairs of uncoupled 2-DoF systems composed of two masses connected through a spring element. The initial guess of the joint FRF $[Y]_o^J$ depends on four parameters: the mass m associated with each translational DoF, the translational stiffness k of the spring that connects two translational DoFs, the inertia I_r associated with each rotational DoFs, and the rotational stiffness k_r that connects two pairs of rotational DoFs.

11.5 Results

In this section, the results of the identification procedure applied to the Brake Reuss Beam are discussed. It is necessary to point out that the results of the identification procedure depend either on the number and the position of measurement and excitation points considered in the generation of the overlay model and on the mass m used for the initial guess of the joint. These dependencies are discussed along with the results obtained in the case in which a sufficient number of experimental measurements and a good estimate of the mass value are used. Since the real behaviour of the joint is not known, the quality of the solution is evaluated by reconstructing the FRF of the parent model $[Y]^{AJB,par}$, obtained by coupling the SEMM models of beams A and B with the identified joint J, in one DoF not involved in the identification process, thus belonging to the set $\{u_o\}^{AB}$. The comparison between this parent model $[Y]^{AJB,par}$ with the FRF of the experimental model at the same DoF provides meaningful information regarding the quality of the solution.

11.5.1 Effect of Measurement Points

The joint identification procedure used in this work is an inverse problem, thus the data space and the solution space, i.e. the dimensions of the overlay model and of the joint substructure, respectively, must be carefully chosen. As seen in Sect. 11.4.1, the joint substructure is defined on a set of 60 DoFs, thus resulting in a FRF matrix $[Y]^J$ with dimension (60×60). The overlay model $[Y]^{AB,ov}$ instead, is defined on N_c measurement DoFs $\{u\}^{AB}$ and on N_e excitation DoFs $\{u_e\}^{AB}$, thus resulting in a matrix with dimension $(N_c \times N_e)$. In this section, the dependence of the solution on the choice of measurement points is investigated. First, the dependence of the solution on the position of the measured DoFs is discussed. In Fig. 11.7, the results of the identification procedure are shown for two different sets of measurement points S_1 and S_2. They have an equal number of measurement and excitation DoF ($N_c = N_e = 25$). In Fig. 11.7a, the joint FRF between the DoFs q_3z and q_8z, obtained for S_1 (blue line) and S_2 (orange line) are shown, while in Fig. 11.7b the corresponding reconstructed FRFs at validation DoF O_2z ($O_2z \in \{u\}_o^A$) and the FRF of the experimental model are compared. Note that the validation DoF O_2z is part of the experimental model, but is kept out of the overlay model, i.e. it is not used in the identification procedure. It can be seen from Fig. 11.7a that in the frequency band from 0 Hz to about 250 Hz, the two FRFs of the joint are quite in agreement and both are able to reconstruct the behaviour of the Brake Reuss Beam in Fig. 11.7b. From 250 Hz on, the FRFs of the joint present a relevant scattering and consequently the corresponding reconstructed FRFs cannot be considered as reliable. The difference in solutions obtained with two different sets of DoFs could be due to ill-conditioning. In particular, the problem could be underdetermined, i.e. the used measurements are not sufficient to reconstruct the real dynamics of the joint. As stated before,

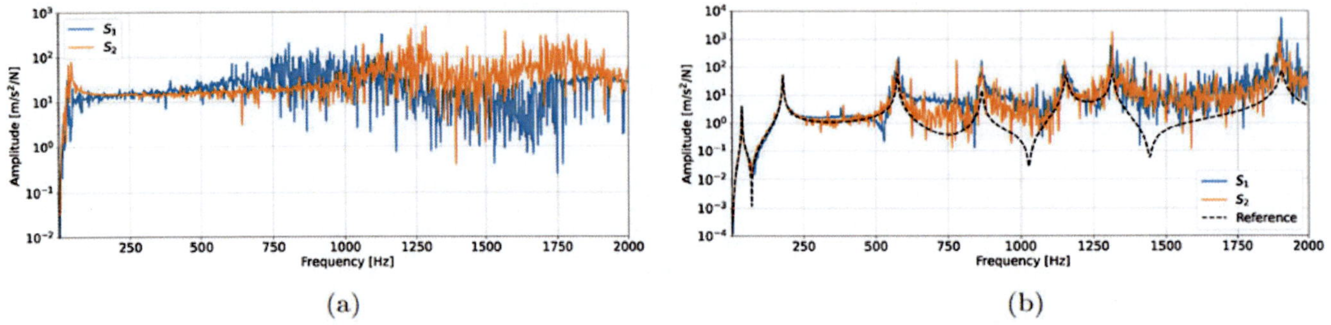

Fig. 11.7 Results with $N_c = N_e = 25$; the blue and orange curves refer to different sets of DoFs, S_1 and S_2, respectively; the black reference curve in (**b**) is obtained from the simulated experimental model. (**a**) Joint FRF between DoFs q_3z and q_8z. (**b**) FRF of the brake reuss beam between the DoFs O_2z on beam A and E_1z on beam B

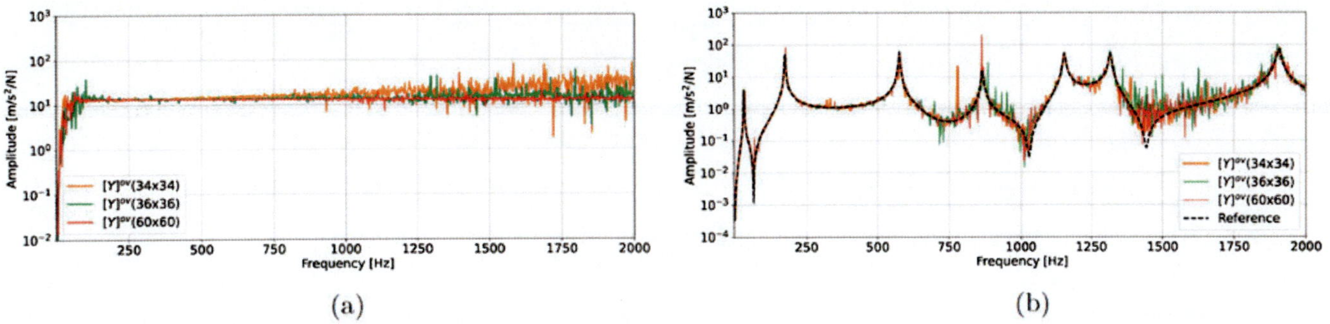

Fig. 11.8 Solution of the identification problem with different dimension of $[Y]^{AB,ov}$: (34 × 34) (orange curve), (36 × 36) (green curve) and (60 × 60) (red curve); the reference curve in black is obtained from the simulated experimental model. (**a**) Joint FRF between DoFs q_3z and q_8z. (**b**) FRF of $[Y]^{AJB,par}$ between DoFs O_2z on beam A and E_1z on beam B

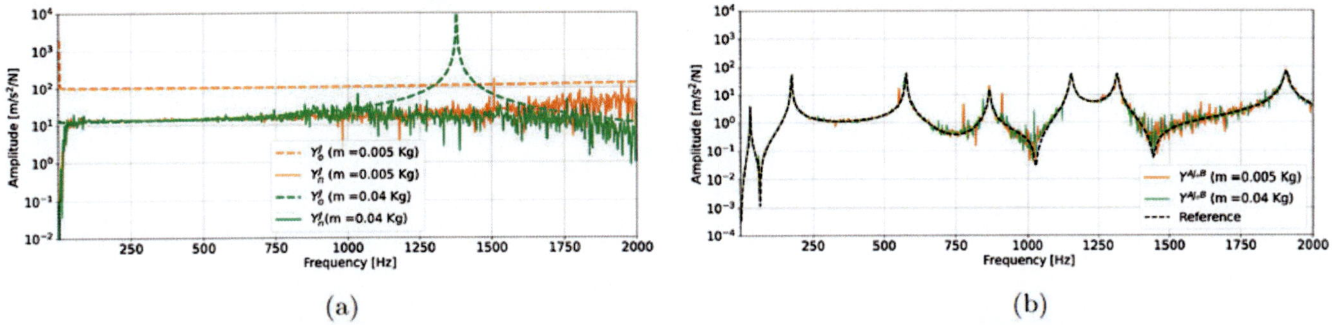

Fig. 11.9 Solution of the identification problem with $[Y]^{AB,ov}(34 \times 34)$, for different values of the initial guess mass m; initial guess FRFs (dashed line); identified FRFs (solid line); the reference curve in black is obtained from the simulated experimental model. (**a**) Joint FRF between DoFs q_3z and q_8z. (**b**) FRF of $[Y]^{AJB,par}$ between the DoFs O_2z on beam A and E_1z on beam B

the joint FRF is defined on a set of 60 DoFs. Thus, by defining the overlay model on a number $Nc = 60$ of measurement DoFs and $N_e = 60$ of excitation DoFs, the problem is determined. The corresponding identified joint FRF is represented by the red curve in Fig. 11.8 that presents a lower scattering with respect to the previous results. Also, the reconstruction of the behaviour of the Brake Reuss Beam (the FRF of the validation point O_2z) is in good agreement with the reference curve (dashed black line) up to 1300 Hz. However, it could be possible that not all the DoFs of the joint substructure are necessary. To investigate this aspect, a lower number of measurement DoFs is considered. In Fig. 11.8, the orange curve refers to $[Y]^{AB,ov}(34 \times 34)$ and the green curve refers to $[Y]^{AB,ov}(36 \times 36)$. In both cases, the reduction of the number of measured DoFs introduces scattering in the joint FRF (Fig. 11.8a) that propagates in the reconstructed FRF of the assembled structure (Fig. 11.8b). However, by comparing the results in Figs. 11.7 and 11.8, it can be noted that the solution improves as the number of measurement DoFs increases.

11.5.2 Effect of Initial Guess Mass Value

In order to start the joint identification procedure, an initial guess model of the connection is required. It is expected that the initial guess value does not affect the final solution and this is verified for the case in which the overlay model has dimension (60×60). However, when the dimension of the overlay model decreases the choice of the initial guess can affect the solution. Therefore, to evaluate the effect of different initial guess model on the solution, the mass value m is increased from 0.005 Kg to 0.04 Kg. In Fig. 11.9, the results of the joint identification carried out with the overlay model $[Y]^{AB,ov}(34 \times 34)$ are shown for two choice of the initial guess mass: $m = 0.005$ Kg and $m = 0.04$ Kg. The two solutions are different, even if they are able to reconstruct the behaviour of the reference model. In particular, it can be noted that above 1500 Hz, where the scattering level becomes more relevant, the joint FRF tends to converge to the initial guess curve. Using an overlay model of dimension (36 × 36), the initial guess change does not affect the solution (Fig. 11.10).

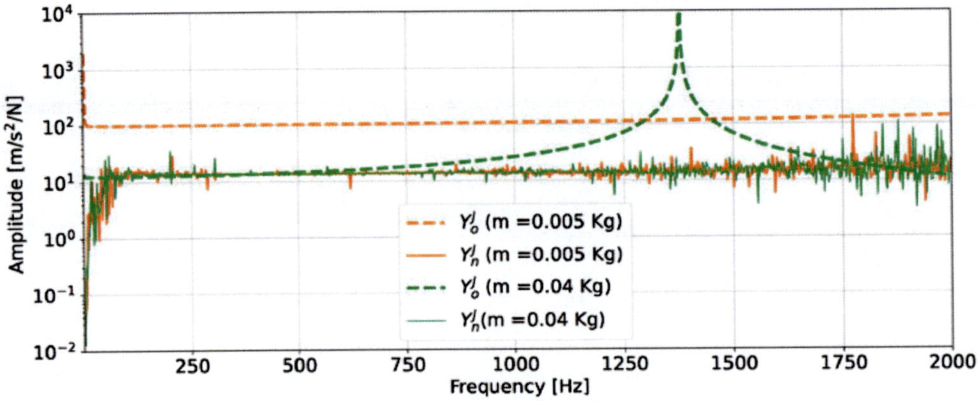

Fig. 11.10 Joint FRF between DoFs q_3z and q_8z with $[Y]^{AB,ov}$(36x36), for different values of the initial guess mass m: initial guess FRFs (dashed line); identified FRFs (solid line)

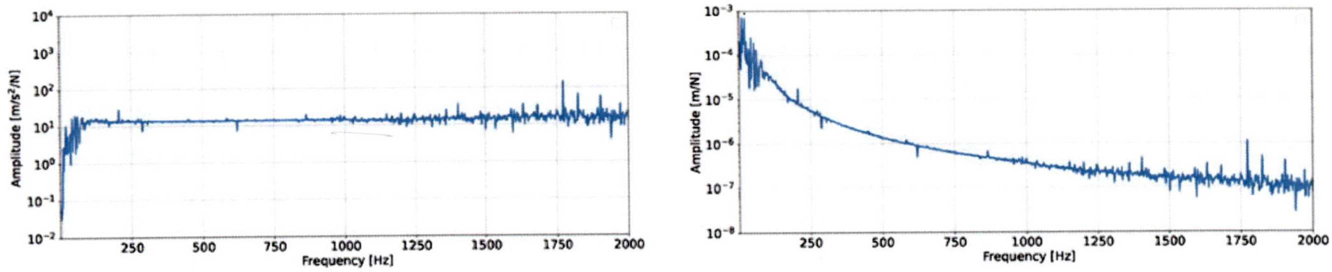

(a) (b)

Fig. 11.11 FRFs of the identified joint between the DoFs q_3z and the q_8z. (**a**) Accelerance FRF. (**b**) Receptance FRF

11.5.3 Identified Joint

The physical interpretation of the identified joint can be carried out by looking at its accelerance and receptance FRFs. The joint identified with the overlay model $[Y]^{AB,ov}(36 \times 36)$ is considered in the following. In Fig. 11.11, the receptance FRF and the accelerance FRFs between the DoFs q_3z and q_8z are shown for the identified joint. These two points are located in correspondence of the central bolt of the Brake Reuss Beam. It can be seen from the receptance FRF that the joint has an initial peak at very low frequency and this is due to the fact the identified joint substructure is in free-free conditions, thus having rigid body modes. At high frequencies, the accelerance of the joint presents a typical mass line.

11.6 Conclusions

In this work, the problem of joint identification is approached by decoupling jointed substructures from the assembled system. In order to obtain information at the coupling DoFs, the System Equivalent Model Mixing expansion has been used and the interface is described using remote points thus taking into account the deformation of the connecting surfaces. The proposed approach has been applied to identify the bolted joint of a numerically simulated Brake Reuss Beam. The obtained results highlight that the choice of the number and location of the measurement DoFs affects the quality of the identification. In fact, by using a set of measurement that makes the problem determined, the quality of the joint identification is quite satisfactory. The set of measurement can be reduced providing acceptable result in a lower frequency band. Conversely, an excessive reduction of the number of measurement DoFs leads to numerical ill-conditioning of the problem due to the lack of relevant information to correctly identify the dynamic behaviour of the joint. Future developments aim at investigating the effect of measurement noise and at applying the identification procedure to different kind of joints using also experimental data.

Acknowledgments This research is supported by University of Rome La Sapienza and University of L'Aquila.

References

1. Brake, M.R.W.: The Mechanics of Joined Structures. Springer, Berlin (2018)
2. de Klerk, D., Rixen, D.J., Voormeeren, S.N.: General framework for dynamic substructuring: history, review and classification of techniques. AIAA J. 46(5), 1169–1181 (2008)
3. Brunetti, J., Culla, A., D'Ambrogio, W., Fregolent, A.: Experimental dynamic substructuring of the Ampair wind turbine test bed. In: Dynamics of Coupled Structures. Proceedings of the Society for Experimental Mechanics Series, vol. 1, pp. 15–26. Springer New York LLC (2014)
4. Klaassen, S., van der Seijs, M., de Klerk, D.: System equivalent model mxing. Mech. Syst. Signal Process. **105**, 90–112 (2018)
5. Jetmundsen, B., Bielawa, R.L., Flannelly, W.G.: Generalized frequency domain substructure synthesis. J. Am. Helicopter Soc. **33**, 55–64 (1988)
6. de Klerk, D., Rixen, D.J., de Jong, J.: The frequency based substructuring (FBS) method reformulated according to the dual domain decomposition method. In: Proceedings of the 24th International Modal Analysis Conference, A Conference on Structural Dynamics, pp. 1–14 (2006)
7. D'Ambrogio, W., Fregolent, A.: Direct decoupling of substructures using primal and dual formulation. In: Linking Models and Experiments. Conference Proceedings of the Society for Experimental Mechanics Series, vol. 2 (2011)
8. D'Ambrogio, W., Fregolent, A.: Decoupling procedures in the general framework of frequency based substructuring. In: Proceedings of the 27th International Modal Analysis Conference (IMAC XXVII), Orlando, FL (2009)
9. D'Ambrogio, W., Fregolent, A.: The role of interface DoFs in decoupling of substructures based on the dual domain decomposition. Mech. Syst. Signal Process. **24**(7), 2035–2048 (2010)
10. Voormeeren, S., Rixen, D.: A dual approach to substructure decoupling techniques. In: Structural Dynamics, vol. 3, pp. 601–616. Springer, Berlin (2011)
11. Voormeeren, S.N., Rixen, D.J.: A family of substructure decoupling techniques based on a dual assembly approach. Mech. Syst. Signal Process. **27**, 379–396 (2012)
12. Haeussler, M., Klaassen, S., Rixen, D.: Comparison of substructuring techniques for experimental identification of rubber isolators dynamic properties. In: ISMA 2018 International Conference on Noise and Vibration Engineering (2018)
13. Haeussler, M., Klaassen, S., Rixen, D.: Experimental twelve degree of freedom rubber isolator models for use in substructuring assemblies. J. Sound Vib. **474**, 115253 (2020)
14. Klaassen, S., Ciğeroğlu, E., Özgüven, H.N.: Identifying the joint dynamics of a bolted beam system by using FRF decoupling and the SEMM expansion technique. In: Proceedings of Tribomechadynamics, Huston, Texas (2019)
15. Saeed, Z., Klaassen, S.W.B., Firrone, C.M., Berruti, T.M., Rixen, D.J.: Experimental joint identification using system equivalent model mixing in a bladed-disk. J. Vib. Acoust. **142**, 1–29 (2020)
16. Klaassen, S.W., Rixen, D.J.: Using SEMM to identify the joint dynamics in multiple degrees of freedom without measuring interfaces. In: Dynamic Substructures. Conference Proceedings of the Society for Experimental Mechanics Series, vol. 4, pp. 87–99. Springer, Berlin (2020)
17. van der Seijs, M., van den Bosch, D.D., Rixen, D.J., de Klerk, D.: An improved methodology for the virtual point transformation of measured frequency response functions in dynamic substructuring. In: 4th ECCOMAS Thematic Conference on Computational Methods in Structural Dynamics and Earthquake Engineering (2013)
18. Ansys, Mechanical APDL Element Reference. ANSYS Inc., Canonsburg, Release R2 (2021)
19. Ansys, Mechanical APDL Theory Reference, ANSYS Inc.,Canonsburg, Release R2 (2021)
20. Bregar, T., El Mahmoudi, A., Kodric, M., Cepon, G., Boltezar, M., Rixen, D.J.: Introducing pyFBS: an open-source python package for frequency based substructuring and transfer path analysis. In: Dynamic Substructures, vol. 4, pp. 81–89. Springer International Publishing (2022)

Chapter 12
Parametric Analysis of the Expansion Process Based on System Equivalent Model Mixing

Miha Kodrič, Tomaž Bregar, Gregor Čepon, and Miha Boltežar

Abstract Accurate determination of the high-resolution dynamic response is a crucial step for the dynamic properties characterization in the development phase of a modern product. The conventional experimental procedure for the identification of the structure's dynamic properties is an experimental modal analysis, where a high spatial resolution can be obtained by imposing a large number of response and excitation points at a structure. To avoid time-consuming experimental testing, measurements involving only a limited number of points at the structure can be expanded to the unmeasured points through model-based expansion techniques. They rely on the introduction of a numerical model, presenting the basics of the expansion. Recently, an expansion method called System Equivalent Model Mixing (SEMM) was proposed where a numerical DoF set is used to extend an experimental model consisting only of a limited number of measurement points. Using the dynamic substructuring approach, the equivalent experimental and numerical models are coupled so that the hybrid model inherits the dynamic properties of both. Although the method has been well adopted, there is still no comprehensive phenomenological analysis to determine the influence of the method's parameters on the consistency of the hybrid model and thus on the accuracy of the expansion process.

This paper addresses the issue by evaluating the accuracy of the SEMM expansion process, focusing on the influence of the regularity of the so-called equivalent numerical model. The introduction of quasi-equivalent numerical models into SEMM is analysed here, which can differ not only with respect to the mass and stiffness properties but also in terms of the geometry and boundary conditions. The parametric study was carried out on a real component of a household appliance, and the most influential parameters in terms of accuracy of the SEMM expansion process were identified.

Keywords Experimental model · Full-field response · Expansion process · System equivalent model mixing · Frequency domain

12.1 Introduction

Consistent dynamic models are a vital tool in the process of designing and optimizing the vibro-acoustic properties of complex products. A conventional way to obtain a dynamic response with a high spatial resolution at a structure is to perform an experimental modal analysis using many impact locations and measuring responses with accelerometers or laser vibrometers. More advanced experimental techniques to acquire dense system response involves laser scanning vibrometer or optical methods, such as Digital Image Correlation and Digital Photogrammetry. Alternatives to determining the dynamic properties in unmeasured degrees of freedom (DoFs) are the model-updating and expansion methods. All model-updating methods are usually very time-consuming and involve computationally intensive operations. On the other hand, expansion methods enable the expansion of the measurements at a limited number of points to a denser set of DoFs defined by the numerical model. Since the measured data are usually in the form of frequency-response functions (FRFs), Klassen et al. [1] introduced an expansion technique called System Equivalent Model Mixing (SEMM), fully defined in the frequency domain. With SEMM, equivalent numerical and experimental models of the same structure are dynamically coupled using the Dynamic Substructuring approach [2].

M. Kodrič (✉) · G. Čepon · M. Boltežar
University of Ljubljana, Faculty of Mechanical Engineering, Ljubljana, Slovenia
e-mail: miha.kodric@fs.uni-lj.si

T. Bregar
Gorenje d.o.o., Velenje, Slovenia

M. Allen et al. (eds.), *Dynamic Substructures, Volume 4*, Conference Proceedings of the Society for Experimental Mechanics Series,
https://doi.org/10.1007/978-3-031-04094-8_12

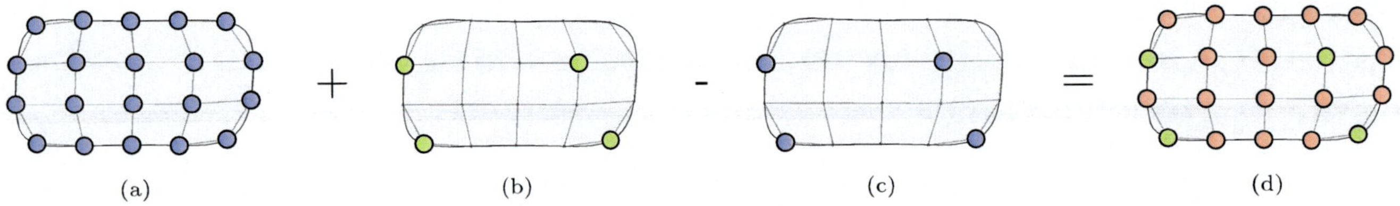

Fig. 12.1 Equivalent models; (**a**) Parent model $\mathbf{Y}^{\mathrm{par}}$, (**b**) Overlay model $\mathbf{Y}^{\mathrm{ov}}$, (**c**) Removed model $\mathbf{Y}^{\mathrm{rem}}$, (**d**) Hybrid model $\mathbf{Y}^{\mathrm{SEMM}}$

This paper evaluates the accuracy of the SEMM expansion process in terms of the regularity of the so-called equivalent numerical model. A quasi-equivalent numerical model is introduced to the SEMM method for which the effect of boundary conditions and geometry simplifications are analysed with regard to the expansion process consistency. In addition, the variation of the experimental model is also analysed with respect to the number and distribution of the excitation and response points. The parametric study is performed on a real-life component of a household appliance to identify the most influential parameters of the SEMM expansion process.

12.2 System Equivalent Model Mixing: SEMM

System Equivalent Model Mixing (SEMM) was introduced by Klaassen et al. [1] and forms a hybrid model by mixing the numerical and experimental FRFs. The method is based on the parent model (Fig. 12.1a), which provides the extensive DoF set and is usually of numerical nature. The dynamic properties are introduced by the overlay model (Fig. 12.1b) obtained by experiment. To form the final hybrid model (Fig. 12.1d) the dynamic properties of the parent model are eliminated with the removed model (Fig. 12.1c). The compatibility and equilibrium conditions between equivalent models at the interface DoFs are imposed using the Lagrange Multiplier Frequency Based Substructuring (LM FBS) framework [2]. Hence, the basic SEMM method takes the following form:

$$\mathbf{Y}^{\mathrm{SEMM}} = \mathbf{Y}^{\mathrm{par}} - \begin{bmatrix} \mathbf{Y}_{\mathrm{ib}} \\ \mathbf{Y}_{\mathrm{bb}} \end{bmatrix}^{\mathrm{par}} \left(\mathbf{Y}^{\mathrm{rem}}\right)^{-1} \left(\mathbf{Y}^{\mathrm{rem}} - \mathbf{Y}^{\mathrm{ov}}\right) \left(\mathbf{Y}^{\mathrm{rem}}\right)^{-1} \left[\mathbf{Y}_{\mathrm{bi}}\ \mathbf{Y}_{\mathrm{bb}}\right]^{\mathrm{par}}. \tag{12.1}$$

The basic SEMM method also has some extensions to increase its robustness, e.g. to remove spurious peaks from the hybrid model. This is achieved by extending the removed interface to all the internal DoFs ($\mathbf{Y}^{\mathrm{par}} = \mathbf{Y}^{\mathrm{rem}}$). The final form of the fully extended SEMM method in a single-line notation is

$$\mathbf{Y}^{\mathrm{SEMM}} = \mathbf{Y}^{\mathrm{par}} - \mathbf{Y}^{\mathrm{par}} \left(\left[\mathbf{Y}_{\mathrm{bi}}\ \mathbf{Y}_{\mathrm{bb}}\right]^{\mathrm{rem}}\right)^{+} \left(\mathbf{Y}_{\mathrm{bb}}^{\mathrm{rem}} - \mathbf{Y}^{\mathrm{ov}}\right) \left(\begin{bmatrix} \mathbf{Y}_{\mathrm{ib}} \\ \mathbf{Y}_{\mathrm{bb}} \end{bmatrix}^{\mathrm{rem}}\right)^{+} \mathbf{Y}^{\mathrm{par}}. \tag{12.2}$$

12.3 Using SEMM as an Expansion Technique

Conventionally, the SEMM method is intended to couple equivalent models of the same structure, whereby the numerical model is considered as a digital twin of the experimental model [1]. As we are often interested in the dynamic response of the certain sub-competent of the assembly only, we propose a different methodology to the SEMM formulation. Here, the experimental model is acquired on a real sub-component, mounted in the assembled configuration, while the numerical model represents the analysed structure isolated from the rest of the system. Although numerical and experimental models are consequently of the same structure, they can be considered as quasi-equivalent since they can differ in terms of geometry and applied boundary conditions. The whole idea is shown schematically in Fig. 12.2.

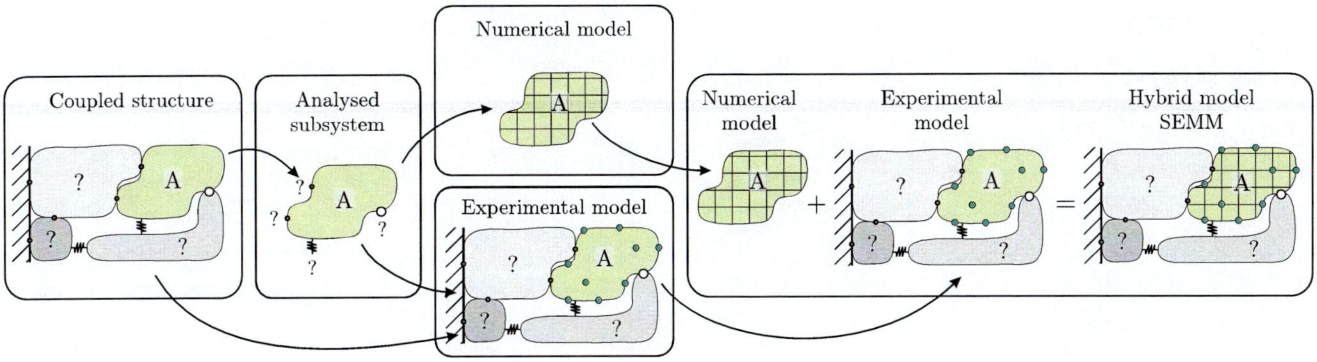

Fig. 12.2 SEMM expansion process including quasi-equivalent models

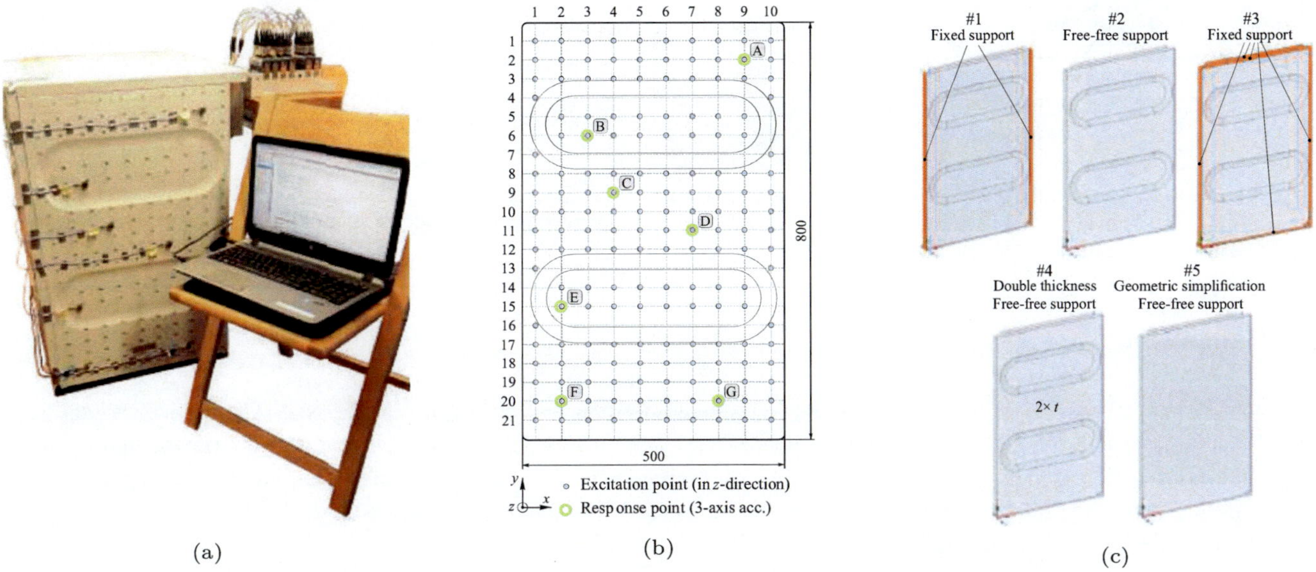

Fig. 12.3 Analysed washing machine; (**a**) experimental setup, (**c**) locations of response and excitation points, (**c**) numerical models

12.4 Analysis of the Method on an Experimental Case Study

The proposed SEMM expansion method was used to expand the vibration response from a limited number of measured points on the side panel of a washing machine (Fig. 12.3a). Performance of the SEMM expansion process was analysed regarding the number and location of the excitation and response points in the experimental model (Fig. 12.3b), as well as with respect to the consistency of the numerical model. Therefore, five different numerical models (Fig. 12.3c) were proposed in the expansion process, which differs in terms of geometry and boundary conditions. The quality of the expansion process was accessed using the FRAC criterion, comparing the expanded full-field vibration response with the actual reference measured response. The distribution of excitation points over analysed surface is quantified using aggregation index R ($R \approx 1$—random distribution, $R \approx 0$—clustering, $R \approx 2$—well-distributed points) [3].

Based on the presentation in Fig. 12.4 it is clear that the accuracy of the expanded displacement field improves by increasing the number of excitation and response points. The most accurate expansion process is obtained using numerical models #1, #2 and #4, as their geometry and boundary conditions are similar to the actual ones (Fig. 12.4a). The expansion processes using numerical models #3 and #5 result in poor quality of the hybrid model since the boundary conditions are too rigid, or the geometry is much too simplified, respectively. Furthermore, a small number of excitation and response points can lead to an inconsistent reconstruction of the vibration response (Fig. 12.4b) since the experimental dynamic model is incomplete (can be related to the poor observability or controllability of the interface DoF). However, even if a small number of excitation points are well scattered over the analysed surface (high aggregation index), these configurations can result in an accurate expansion process (Fig. 12.4c).

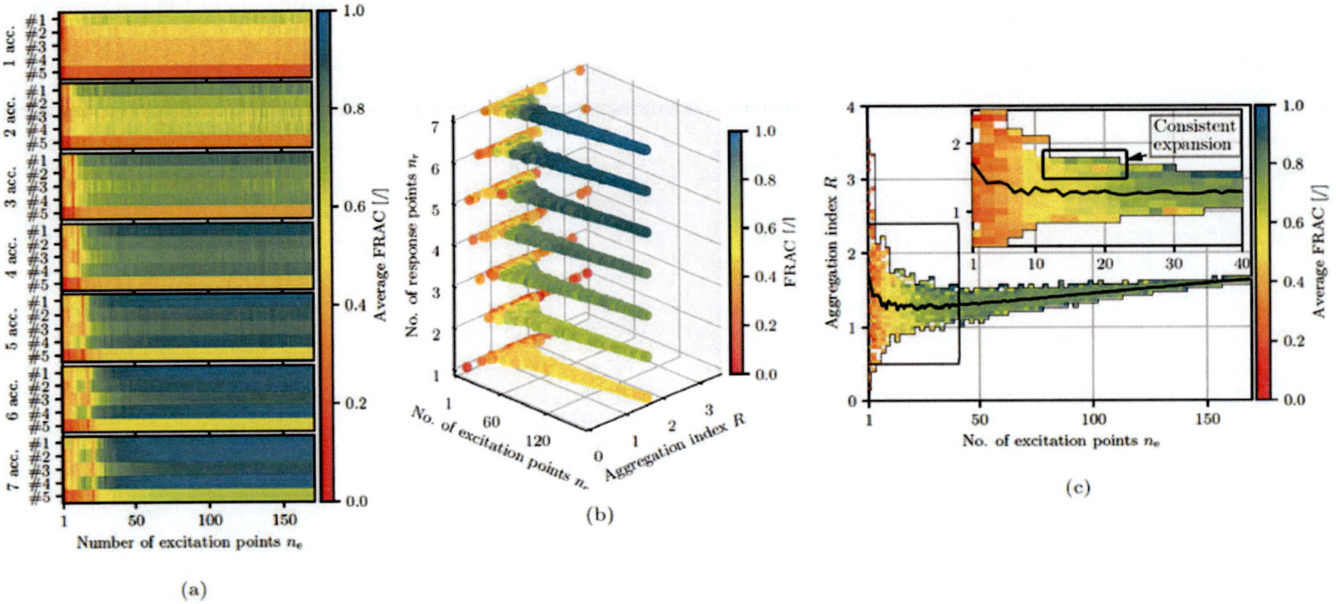

Fig. 12.4 Results of the parametric analysis using FRAC criterion; (**a**) influence of number of response and excitation points considering different numerical models, (**b**) detailed results of the analysis using numerical model #1 as parent model, (**c**) influence of the aggregation index R and the number of excitation points in the experimental model

12.5 Conclusion

This study provides a comprehensive analysis of the parameters influencing the quality of the SEMM expansion process. Using a quasi-equivalent numerical model, it was shown that the expansion process is more accurate the more excitation and response points are included in the experimental model. An accurate expansion process with a small number of response and excitation points is feasible only when the experimentally observed DoFs are well scattered over the given object and the numerical model presents consistent basics for the expansion process.

References

1. Klaassen, S.W., van der Seijs, M.V., de Klerk, D.: System equivalent model mixing. Mech. Syst. Signal Process. **105**, 90–112 (2018)
2. de Klerk, D., Rixen, D.J., de Jong, J.: The frequency based substructuring (FBS) method reformulated according to the dual domain decomposition method. In: Proceedings of the 24th International Modal Analysis Conference, A Conference on Structural Dynamics, pp. 1–14 (2006)
3. C.J. Krebs, Ecological Methodology. The Benjamin/Cummings Series in the Life Sciences. Benjamin/Cummings, San Francisco (1999)

Chapter 13
Real-Time Hybrid Simulation Study of a Physical Duffing Absorber Attached to a Virtual Nonlinear Structure

A. Mario Puhwein and Markus J. Hochrainer

Abstract In structural dynamics, real-time hybrid simulation is used for physical testing of critical components. When compared to traditional testing methods, the approach is very efficient and reliable, if important surrounding effects can be included in a virtual model (simulation model) which is interacting with the physical component (physical model). The virtual model must be processed in real time, and proper coupling is essential for reliable and consistent test results. Consequently, the physical interface conditions must be satisfied, and therefore, a transfer system is generally required for component testing. In the work presented, a real-time hybrid simulation experiment is used to study the influence of a Duffing absorber on the dynamics of the fundamental mode of vibration of a nonlinear host structure. Only standard hardware components are used, e.g., electrodynamic shaker, force, displacement and acceleration sensors, as well as a real-time processing unit. The study is currently limited to periodic forcing, and consequently, only the steady-state response is studied. Since the physical absorber investigated behaves nonlinear elastic, all measurements can be collected sequentially to generate high-precision force response diagrams for different amplitudes and frequencies. So far, the experimental results agree well with theoretical predictions and confirm that nonlinear dynamic absorbers are particularly suited to mitigate vibrations of nonlinear host structures. As the properties of the nonlinear absorber are adjustable, existing design rules can be verified by the proposed experimental setup.

Keywords Experimental nonlinear vibrations · Periodic forcing · Duffing oscillator · Magnetic forces · Force displacement hysteresis

13.1 Introduction

Real-time hybrid simulation (RTHS) also known as real-time hybrid testing (RTHT), cyberphysical simulation or testing, as well as hardware-in-the-loop (HIL) testing has become increasingly popular in recent years. This experimental technique allows testing of physical components or modules under almost realistic conditions, because it takes into account the influence of the adjacent virtual components, which are simulated in a real-time computer environment. The testing methodology has become widespread because of major improvements in modern simulation techniques, the availability of real-time systems, and significant advances in the simulation and analysis of hybrid systems.

The real-time simulation model is commonly denoted RTHS simulator, and it is coupled to the physical model via appropriate interfaces. The entire experiment consisting of RTHS simulator, interfaces, and physical model is called RTHS test bench. Due to technological advances in recent years, it is now possible to test not only individual components but also more complex units. To specify the respective HIL application more precisely, the test bench function is directly included in the name, for example, "steering in the loop," "braking in the loop," or "vehicle in the loop" in vehicle development or "absorber in the loop" in the field of vibration damping.

The development of the RTHS simulator usually requires complex virtual models in different physical domains, whose development is supported by the concept of model-based design. Modern simulation development environments combine elements of mathematical and physical modelling with graphical programming and support automated code generation

A. M. Puhwein
Technical University of Vienna, Vienna, Austria

M. J. Hochrainer (✉)
University of Applied Sciences, Wiener Neustadt, Austria
e-mail: markus.hochrainer@fhwn.ac.at

M. Allen et al. (eds.), *Dynamic Substructures, Volume 4*, Conference Proceedings of the Society for Experimental Mechanics Series,
https://doi.org/10.1007/978-3-031-04094-8_13

for fast development cycles. There are several commercial simulation tools available that support engineers in modeling, analyzing, and optimizing multiphysic coupled systems. Usually, they also provide high-performance mechanical, electrical, or fluid dynamic elements in highly specialized libraries, thus enabling the fast development of complex simulations [1, 2]. The standardization of functional mockup units (FMU) and functional mockup interfaces (FMI) can be used to develop important parts of the virtual models individually in their respective domain with different tools and link them together to the simulation model.

In almost all RTHS applications, the mathematical representation of the simulation model is independent of the real-time target hardware used. Development and simulation environments support automated code generation for hardware components of different performance levels, e.g., embedded systems, PC-based real-time systems, or FPGA platforms. Since the physical model is available, the reliability of the RTHS experiment depends on two factors: the quality of the simulation model and the correct coupling with the physical system. If the physical components are "cut free" from the overall system, correct coupling demands that the potential and flow conditions are fulfilled. If a transfer system is required for proper coupling, it will limit the dynamic range of the RTHS experiment. Because of the simple basic principle, the great flexibility, and the efficiency, the RTHS methodology has been developed independently in different disciplines.

RTHS experiments can test components in the normal operation range, but also in conditions that are hardly possible with conventional experiments, including overloading, usage in an unintended amplitude or frequency range, feeding in error signals, modification of sensor or actuator properties (aging, drift, noise), failure of individual subcomponents, or errors in data transmission. Furthermore, RTHS experiments can be conducted as soon as a physical model is available. This can be long before prototypes of the adjacent systems are available. The relatively easy access and the early availability in the development process are further salient features of RTHS.

13.2 Absorber Testing and Controller Concept

When studying nonlinear vibration absorption with RTHS, the real absorber (physical model) must be separated from the main structure (the virtual model). If the host structure is represented by a plane shear frame structure, the dominant displacement $x_A(t)$ is assumed horizontal; thus, the corresponding interaction force $F_A(t)$ must be horizontal, too. Since the real absorber is attached to its host structure, it has to follow a prescribed motion. In the experiment, this motion is produced by a controlled actuator system commonly denoted "transfer system." In the setup proposed, it consists of a feedback-controlled electrodynamic vibration exciter, a power amplifier, as well as displacement and acceleration sensors. The dynamics of the virtual structural model is simulated with a real-time PC, and analog as well as digital IO components are necessary for the coupling to the physical model. For the configuration of the RTHS experiment, the test sequence automation, data recording, and analysis, an additional host PC is required; see Fig. 13.1, for the interaction of the main components.

Although the use of shaker tests and the RTHS philosophy is well established and used for several decades in experimental vibration analysis, the correct coupling of virtual and experimental model and the required feedback control still represent a

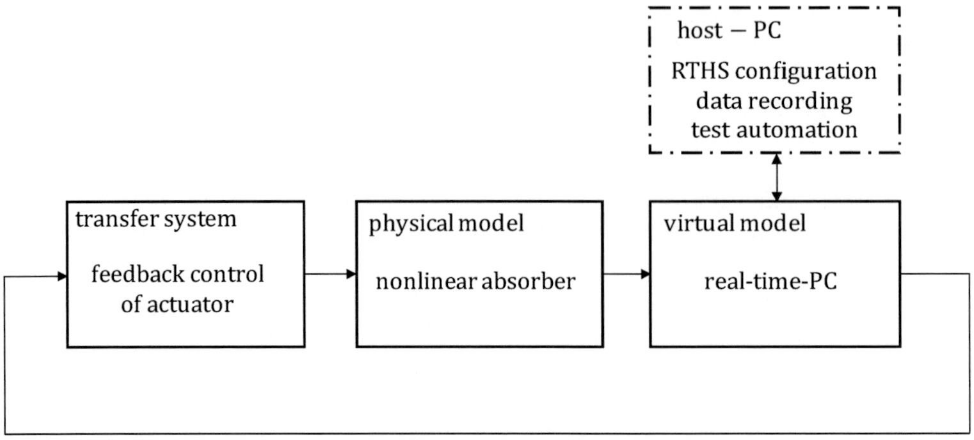

Fig. 13.1 RTHS loop with essential components: transfer system for coupling the submodels; physical model of the nonlinear absorber; real-time PC with virtual model of nonlinear main structure; host PC for control of experiment

major challenge [3–6]. Generally, the dynamics of the device under test is not exactly known, and consequently, the dynamic interaction force between shaker and virtual model remains unknown. Depending on the type of actuator, there are delay times, friction, and many other (nonlinear) effects making the correct coupling even more difficult. Thus, the authors are not aware of a universal and optimal control algorithm. However, if the system behaves nonlinear elastic and periodic forcing is considered, the "iterative learning control" (ILC) or "repetitive control" (RC) [7–9] can be applied very successfully. It can be used in addition to classic feedback control and leads to a flexible and precise adaptive control strategy.

13.3 System Description

In the vibration analysis of complex structures with many degrees of freedom, it is common to identify critical vibration modes and generate a reduced structure model with a few degrees of freedom. In many cases, it is possible to represent the structural oscillation of the host structure (index *S*) by a single dominant oscillation mode of a Duffing-type oscillator

$$m_S\ddot{x}_S + c_S\dot{x}_S + k_S x_S + k_{S3}{x_S}^3 = F\ \cos\omega t, \tag{13.1}$$

with x_S, m_S, c_S, k_S, and k_{S3} denoting the displacement, mass, linear viscous damping, stiffness, and cubic stiffness constants. F and ω describe the amplitude and frequency of the harmonic force excitation. For the subsequent analysis, the equation of motion (13.1) can be given in a dimensionless representation by introducing the normalized time $\tau = \omega_{nS}t$, the natural frequency $\omega_{nS} = \sqrt{k_S/m_S}$, the linear damping coefficient $\zeta_S = c_S/(2m_S\omega_{nS})$, and the nonlinear stiffness ratio $\widetilde{\kappa}_3 = 3k_{S3}/4k_S$. Then normalization of the force excitation renders $f = F/k_S$, $\gamma = \omega/\omega_{nS}$, and $\kappa_3 = \widetilde{\kappa}_3 f^2$. Describing the displacement and the normalized time derivative by $q_S = x_S/f$ and $d/d\tau\ () = ()'$, respectively, finally yields

$$q_S'' + 2\zeta_S q_S' + q_S + \frac{4}{3}\kappa_3 q_S^3 = \cos(\gamma\tau)\,. \tag{13.2}$$

Since there is no exact analytical solution for Eq. 13.2, it is commonly analyzed by the method of harmonic balance. Describing the vibration amplitude by $q_S = A_S \cos\gamma\tau + B_S \sin\gamma\tau$, substituting $\cos^3\gamma\tau \approx 3/4 \cos\gamma\tau$ and $\sin^3\gamma\tau \approx 3/4 \sin\gamma\tau$, and comparing sine and cosine coefficients render two nonlinear algebraic equations for the unknown amplitudes A_S, B_S

$$\begin{aligned} \kappa_3 A_S^3 + \kappa_3 A_S B_S^2 + \left(1-\gamma^2\right) A_S + 2\zeta_S\gamma B_S &= 1, \\ \kappa_3 A_S^2 B_S + \kappa_3 B_S^3 + \left(1-\gamma^2\right) B_S - 2\zeta_S\gamma A_S &= 0. \end{aligned} \tag{13.3}$$

Summing up the squares of Eq. (13.8) and substituting $q_{S,\mathrm{eff}}^2 = A_S^2 + B_S^2$ yield a third-order polynomial equation for $q_{S,\mathrm{eff}}^2$

$$\left[\left(\kappa_3 q_{S,\mathrm{eff}}^2 + 1 - \gamma^2\right)^2 + (2\zeta_S\gamma)^2\right] q_{S,\mathrm{eff}}^2 = 1, \tag{13.4}$$

which must be solved numerically to obtain frequency response functions of the Duffing-type structure; see Fig. 13.2 for an oscillator with a hardening spring and different excitation and damping levels, where $a_F = F/m_S$ denotes the excitation amplitude.

If a vibration absorber is used to improve the dynamic behavior of the main structure at frequencies close to resonance, the system can be approximated by a two-mass oscillator. The use of vibration absorbers for linear systems is well established in engineering; the first work dates back to the beginning of the twentieth century, when Frahm [7] and Den Hartog [8] presented optimal design criteria for systems with narrow and broadband excitation. Recently, the concept of vibration damping has been successfully transferred to nonlinear systems; Habib et al. [9] published criteria for dynamic absorbers with cubic nonlinearity. They have found that the main structure and absorber must have the same nonlinear properties for effective vibration reduction. Accordingly, a Duffing-type main structure requires a Duffing-type absorber for optimal vibration reduction; see Fig. 13.3a for a schematic representation of the system, which also indicates the separation into a virtual and a physical model. The coupling of both models is achieved by the transfer system which has to guarantee that the dynamic equilibrium and compatibility conditions are fulfilled.

In the case of perfect coupling, the mathematical description of the two DOF systems becomes

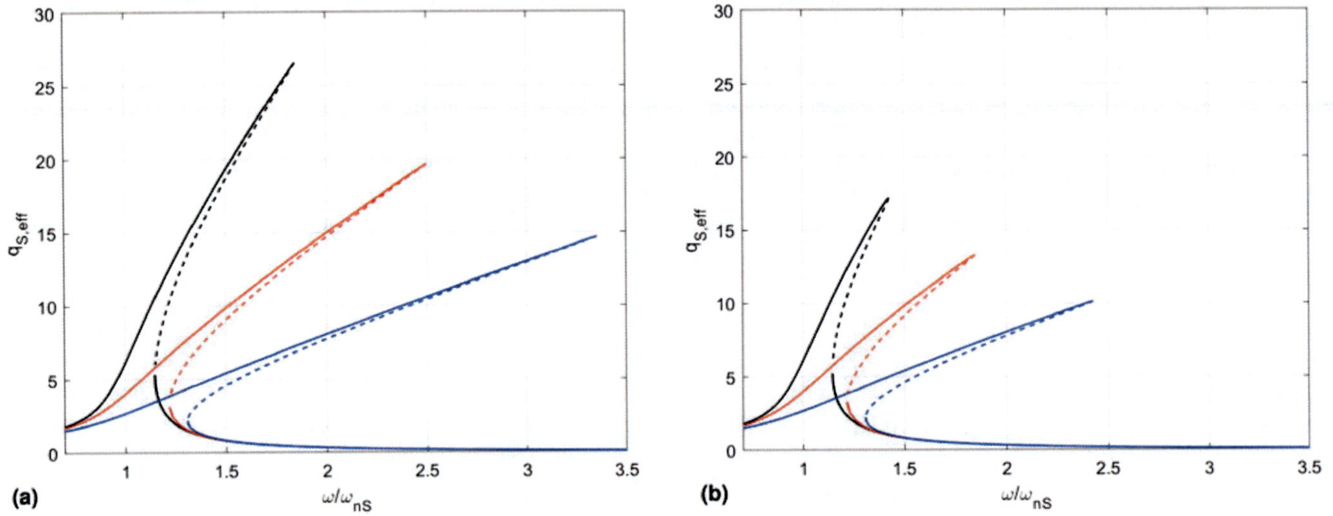

Fig. 13.2 Different frequency response functions of single degree of freedom structures with different damping ratio, — stable and – unstable branch, – $a_F = 0.04$ m/s^2, – $a_F = 0.08$ m/s^2, – $a_F = 0.15$ m/s^2 (**a**) $\zeta_S = 1\%$ (**b**) $\zeta_S = 3\%$

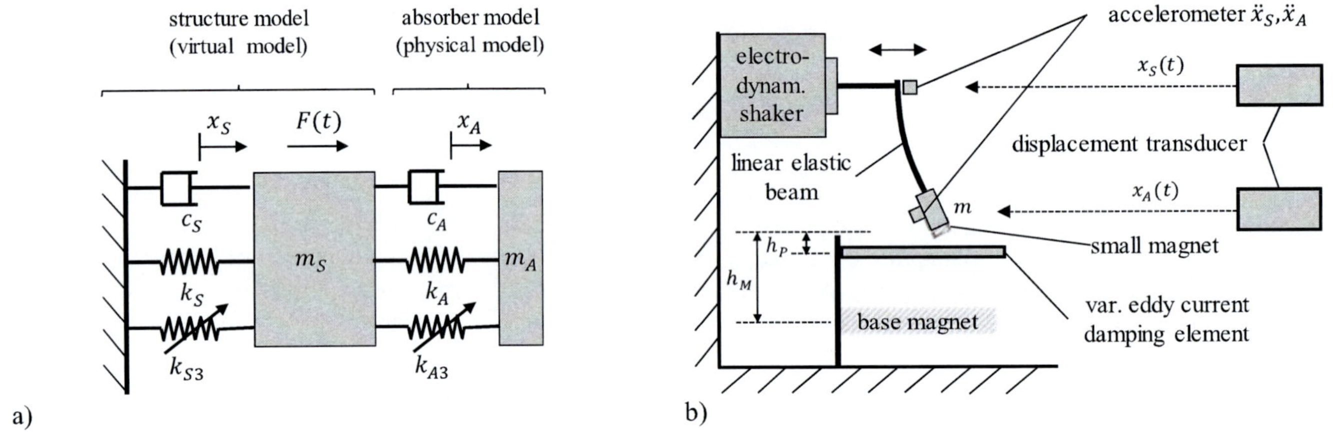

Fig. 13.3 (**a**) Schematic representation of a nonlinear oscillator with a nonlinear dynamic absorber. (**b**) Model of a damper with nonlinear restoring forces caused by repelling magnets in combination with a linear elastic leaf spring

$$\begin{aligned} m_S\ddot{x}_S + c_S\dot{x}_S + k_S x_S + k_{S3}{x_S}^3 &= F\ \cos\omega t - F_A, \\ m_A\ddot{x}_A + c_A\left(\dot{x}_A - \dot{x}_S\right) + k_A\left(x_A - x_S\right) + k_{A3}(x_A - x_S)^3 &= 0, \end{aligned} \tag{13.5}$$

where x_S, m_S, c_S, and k_S describe the absolute displacement, effective mass, damping, and stiffness of the main structure and x_A, m_A, c_A, and k_A represent the corresponding values of the vibration absorber. The absorber-structure interaction force F_A becomes

$$F_A = m_A\ddot{x}_A = -c_A\left(\dot{x}_A - \dot{x}_S\right) - k_A\left(x_A - x_S\right) - k_{A3}(x_A - x_S)^3. \tag{13.6}$$

Again, a dimensionless representation is obtained by introducing the normalized time $\tau = \omega_{nS}t$, the natural frequency $\omega_{nS} = \sqrt{k_S/m_S}$, the structure damping coefficient $\zeta_S = c_S/(2m_S\omega_{nS})$, and the nonlinear stiffness ratio $\widetilde{\kappa}_3 = 3k_{S3}/4k_S$. The absorber natural frequency and linear damping coefficient become $\omega_{nA} = \sqrt{k_A/m_A}$, $\zeta_A = c_A/(2m_A\omega_{nA})$. Given the absorber-structure frequency ratio $\lambda = \omega_{nA}/\omega_{nS}$, the mass ratio $\mu = m_A/m_S$, the equivalent static deformation $x_{S,\,stat} = F/k_S$, the normalized excitation frequency $\gamma = \omega/\omega_{nS}$, as well as the normalized displacements $q_S = x_S/x_{S,\,stat}$, $q_A = (x_S(t) - x_A(t))/x_{S,\,\mathrm{stat}}$, the coupled equations of motion become [9, 10]

$$\begin{aligned} q_S'' + 2\zeta_S q_S' + q_S + \tfrac{4}{3}\kappa_3 q_S^3 + 2\zeta_A\lambda\mu q_A' + \lambda^2\mu q_A + \tfrac{4}{3}\mu\beta_3 q_A^3 &= \cos\left(\gamma\tau\right), \\ q_A'' + 2\zeta_S q_S' + q_S + \tfrac{4}{3}\kappa_3 q_S^3 + 2\zeta_A\lambda\left(\mu + 1\right) q_A' + \lambda^2\left(\mu + 1\right) q_A + \tfrac{4}{3}\left(\mu + 1\right)\beta_3 q_A^3 &= \cos\left(\gamma\tau\right), \end{aligned} \tag{13.7}$$

where

$$\kappa_3 = \tilde{\kappa}_3 f^2, \qquad \beta_3 = \frac{3}{4}\frac{x_{S,stat}^2 k_{A3}}{m_A \omega_{nS}{}^2}. \tag{13.8}$$

Inserting $q_A = A_A \cos \gamma\tau + B_A \sin \gamma\tau$ and $q_S = A_S \cos \gamma\tau + B_S \sin \gamma\tau$ according to the method of harmonic balance, substituting $\cos^3\gamma\tau \approx 3/4 \cos \gamma\tau$ and $\sin^3\gamma\tau \approx 3/4 \sin \gamma\tau$, and comparing sine and cosine coefficients render a set of four nonlinear algebraic equations

$$\begin{gathered}
\kappa_3 A_S{}^3 + \kappa_3 A_S B_S{}^2 + \beta_3 \mu A_A{}^3 + \beta_3 \mu A_A B_A{}^2 + \left(1-\gamma^2\right) A_S + 2\zeta_S \gamma B_S + \lambda^2 \mu A_A + 2\zeta_A \lambda \mu \gamma B_A = 1, \\
\kappa_3 A_S{}^3 + \kappa_3 A_S B_S{}^2 + \beta_3 (\mu+1) A_A{}^3 + \beta_3 (\mu+1) A_A B_A{}^2 + A_S + 2\zeta_S \gamma B_S + \left[\lambda^2 (\mu+1) - \gamma^2\right] A_A \\
+ 2\zeta_A \lambda (\mu+1) \gamma B_A = 1, \\
\kappa_3 A_S{}^2 B_S + \kappa_3 B_S{}^3 + \beta_3 \mu A_A{}^2 B_A + \beta_3 \mu B_A{}^3 - 2\zeta_S \gamma A_S + \left(1-\gamma^2\right) B_S - 2\zeta_A \lambda \mu \gamma A_A + \lambda^2 \mu B_A = 0, \\
\kappa_3 A_S{}^2 B_S + \kappa_3 B_S{}^3 + \beta_3 (\mu+1) A_A{}^2 B_A + \beta_3 (\mu+1) B_A{}^3 - 2\zeta_S \gamma A_S + B_S - 2\zeta_A \lambda (\mu+1) \gamma A_A \\
+ \left[\lambda^2 (\mu+1) - \gamma^2\right] B_A = 0.
\end{gathered} \tag{13.9}$$

To determine the unknown amplitudes A_S, B_S, A_A, B_A, these equations must be solved numerically. In a typical dynamic absorber design procedure, the absorber mass defines the highest possible vibration mitigation and is maximized with respect to the structural limitations. If $\mu = m_A/m_S$ is known, the absorber frequency and damping are determined by the well-established Den Hartog's design criteria $\lambda = 1/(1 + \mu)$ and $\zeta_A = \sqrt{3\mu/(8+8\mu)}$ for linear systems [11]. For nonlinear absorption, there is, furthermore, the cubic stiffness parameter k_{3A} which can also be determined following Den Hartog's equal peak design rules. k_{3A} is optimal, if the maximal structural amplification q_S becomes minimal in the frequency range of interest. This optimization is generally done numerically; however, [9] published the simple approximation

$$k_{A3} \cong \frac{2\mu^2 k_{S3}}{(1+4\mu)}. \tag{13.10}$$

Even if Eq. 13.10 does not render the optimal design for all parameter configurations, it describes a very useful initial condition for numerical optimizations. Figure 13.4a displays the frequency response amplification q_S after Den Hartog's equal peak design method for different excitation levels. Apparently, the absorber performs well for low forcings, whereas for high excitation amplitudes, the nonlinearities dominate the response and prevent the successful application of dynamic absorbers. This effect is shown in Fig. 13.4a for $a_F > 0.12$. In this excitation regime, the absorber response illustrated in Fig. 13.4b remains almost constant, whereas the structural displacements increase dramatically. For $a_F < 0.12$, structure and absorber show an almost optimal performance. The results of Fig. 13.4 were obtained with the following parameters: $m_S = 3.51$ kg, $k_S = 1.6789 \times 10^3$N/m, $\omega_S = 21.87$rad/s, $c_S = 1.5350$ Ns/m, $\zeta_S = 1\%$, and $k_{S3} = 2.1117 \times 10^8$N/m^3. A mass ratio of $\mu = m_A/m_S = 1/40$ was chosen rendering $m_A = 80.5 \times 10^{-3}$kg, $k_A = 39.95$ N/m, $\omega_A = 21.34$rad/s , and $\zeta_A = 7.27\%$, $k_{A3} = 4.0307 \times 10^5$N/m^3.

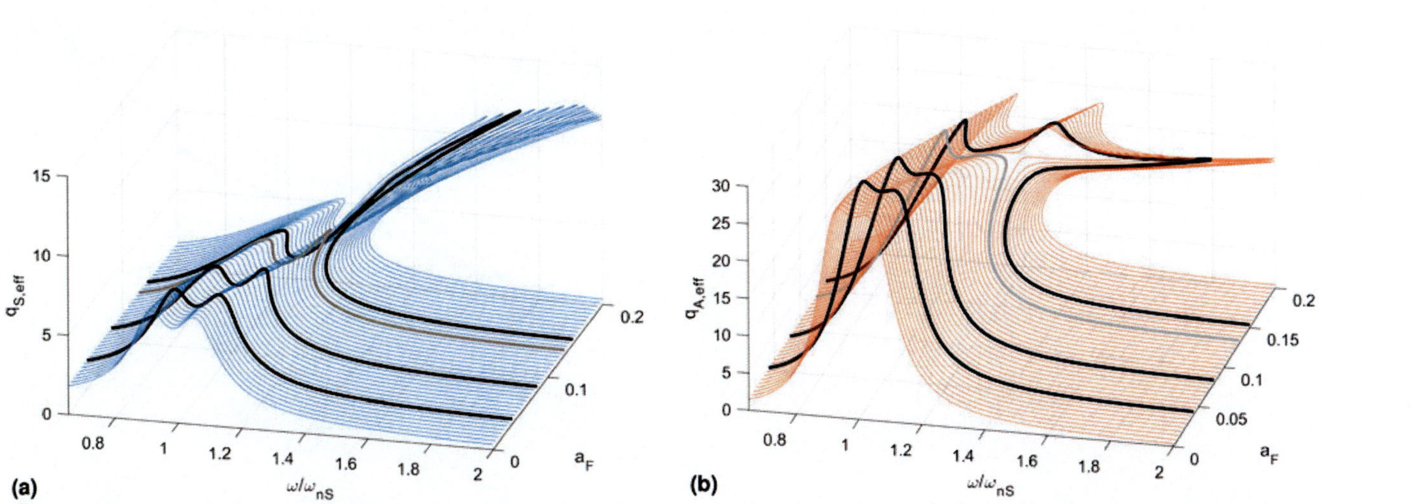

Fig. 13.4 Frequency response surface of nonlinear absorption from harmonic balance analysis: (**a**) structural response $q_{S,\mathrm{eff}} = \sqrt{(A_S^2 + B_S^2)}$ and (**b**) absorber response $q_{A,\mathrm{eff}} = \sqrt{A_A^2 + B_A^2}$

13.4 Experimental Results

After the successful numerical study, experiments are performed to prove the nonlinear dynamic absorption under more realistic conditions. For this purpose, a Duffing-type nonlinear vibration absorber is built and tested in the laboratory. The absorber consists of a beam-type linear elastic spring k_A with a concentrated absorber end mass m_A. In order to generate the required cubic restoring force, a small magnet is attached to the absorber mass. Both are moving in a magnetic field, which is generated by a strong neodymium magnet (base magnet); see Fig. 13.3b. It has been shown that this test arrangement is suitable for experimentally simulating the behavior of the Duffing oscillator [12]. With attracting or repelling magnet configurations, monostable/bistable oscillators with increasing ($k_{A3} > 0$) or decreasing ($k_{A3} < 0$) spring stiffness can be obtained.

To assess the quality of the Duffing absorber, the steady-state force displacement response is measured for two different excitation levels; see Fig. 13.5. The stiffening spring characteristic is evident, but also the hysteretic behavior caused by eddy current damping, which determines the linear viscous absorber damping c_A. The unknown absorber parameters k_A, k_{A3}, and c_A identified from those measurements using least squares show excellent agreement when inserted into Eq. 13.6 and compared with the experiment.

Next, it is demonstrated that the absorber model is able to describe the typical frequency response of a Duffing absorber; see Fig. 13.6. The expected amplitude dependence of the resonance frequency is apparent and results in multiple equilibrium points with different response amplitudes q_A. The "jumping frequency" is depending on whether the excitation frequency is increasing or decreasing. When comparing the FRF of simulated and tested Duffing oscillators, an excellent agreement is found.

Since the physical model of the Duffing absorber behaves as expected, the entire RTHS system can be considered; see Fig. 13.7. In this arrangement, the absorber base acceleration $\ddot{x}_S$ is defined by the corresponding acceleration $\ddot{x}_{S_{\text{ref}}}$ of the simulation model, and the RTHS loop is closed by feeding back the absorber force F_A to the virtual model. The transfer system consists of an electrodynamic shaker connected to a power amplifier and a suitable feedback controller. For an ideal transfer system, shaker and reference acceleration are equal, $\ddot{x}_S = \ddot{x}_{S_{\text{ref}}}$; thus, the transfer function becomes $G = \ddot{x}_S / \ddot{x}_{S_{\text{ref}}} \approx 1$ in the relevant frequency range. The practical limits of the RTHS experiment are mainly determined by the transfer system due to limitations in force, stroke, and frequency range. It has been demonstrated [13] that iterative learning control schemes are perfectly suited to operate the transfer system in steady-state conditions. The entire setup results in cascaded control loops, with the transfer system as inner and the RTHS feedback as outer control loop.

Steady-state experiments are performed in the laboratory for sine stepping excitation with decreasing frequencies; see Fig. 13.8a for the measured structural displacement $x_S(t)$. Although steady-state conditions are reached after several seconds, the frequency remains constant for 30 s before changing the excitation. The settling time depends, among others, on the structural or absorber damping, the (nonlinear) stiffness, as well as the excitation amplitude and frequency. Since the displacement measurements are rather noisy, the steady-state amplitudes are averaged over ten periods. Although it is possible to filter the measured signals before further processing, this would result in an undesired phase delay, which changes or even destabilizes

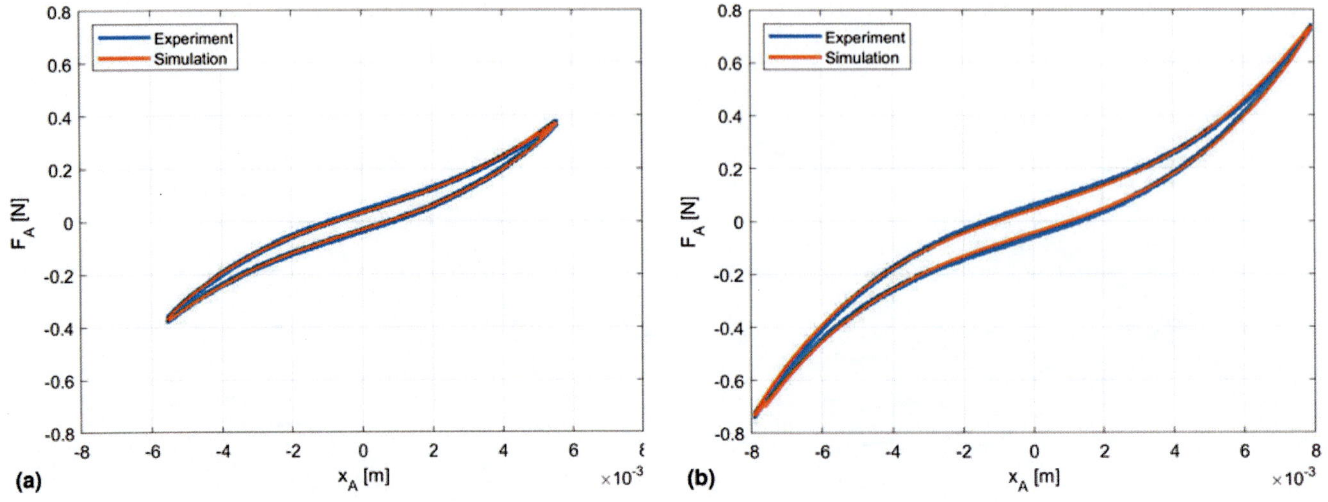

Fig. 13.5 Comparison of measured and simulated force displacement characteristics of Duffing absorber, stiffening configuration: (**a**) steady-state response for $a_F = 0.08$ m/s^2 and (**b**) steady-state response for $a_F = 0.15$ m/s^2

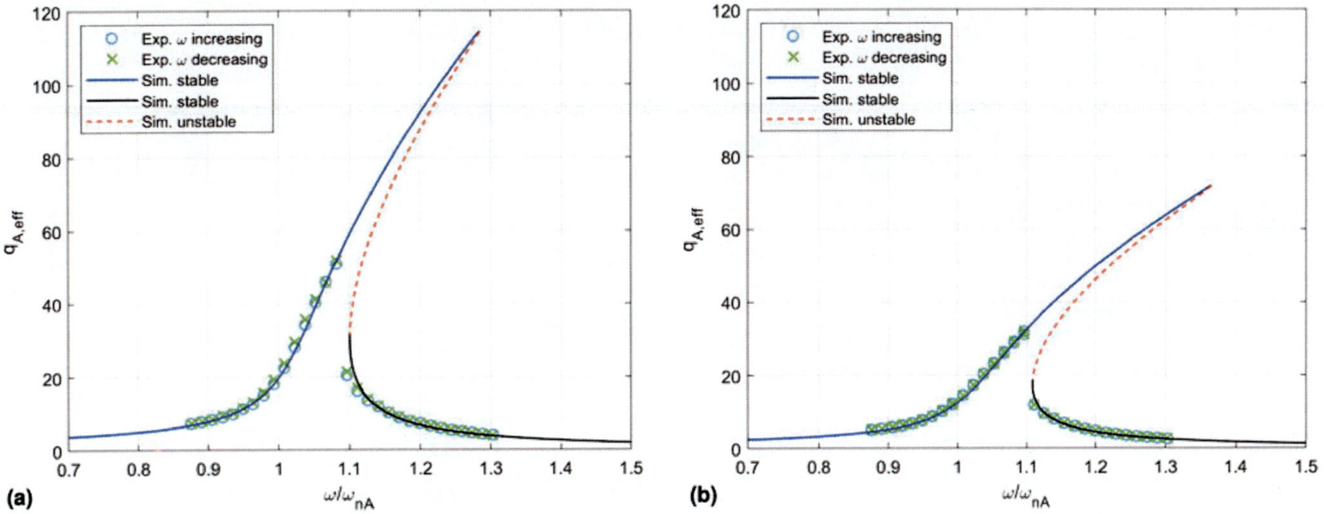

Fig. 13.6 Comparison of measured and simulated FRF of the Duffing absorber for different excitation levels: (**a**) steady-state response for $a_F = 0.08$ m/s^2 and (**b**) steady-state response for $a_F = 0.15$ m/s^2

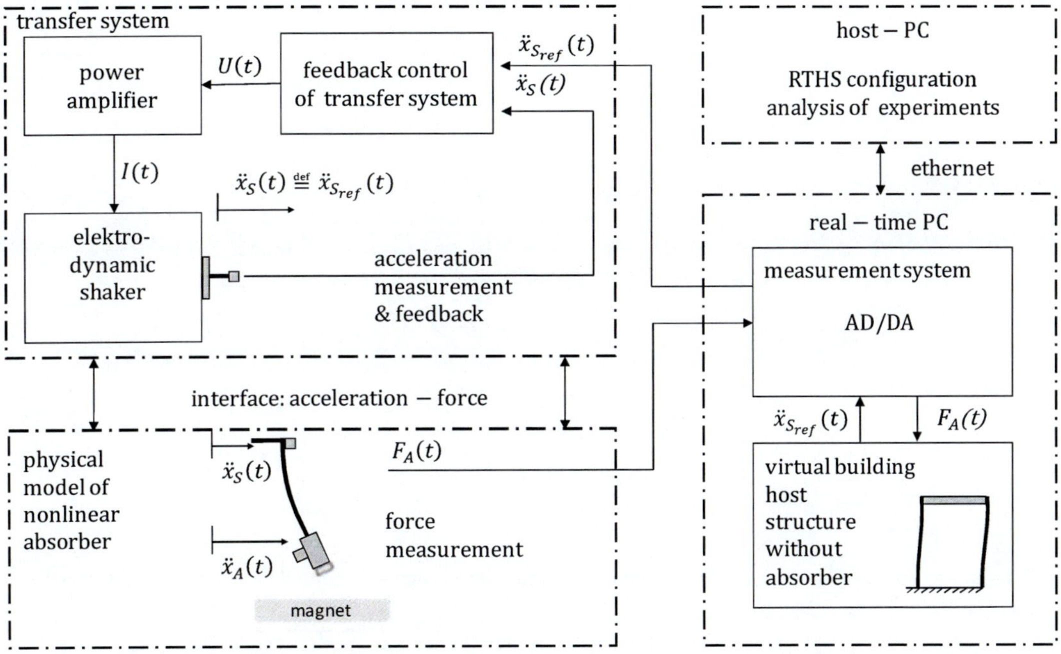

Fig. 13.7 Detailed illustration of RTHS loop with relevant interface quantities

the RTHS experiment. Therefore, filtering is kept minimal, and detailed inspection (see Fig. 13.8b) reveals that x_S perfectly approximates $x_{S\mathrm{ref}}$.

The experiments have been performed in the frequency range between $f_{\min} = 3$ Hz and $f_{\max} = 5$ Hz for excitation levels of $a_F = 0.04$ m/s^2 and $a_F = 0.08$ m/s^2, respectively, while recording the maximum steady-state displacement amplitude $x_{S\,\max}$ of the host structure. Figure 13.9 displays the nonlinear vibration response for an almost perfectly tuned 2DOF in the low amplitude range of $a_F < 0.12$ m/s^2. In Fig. 13.9, the RTHS experiments are compared to the numerical simulation results in Fig. 13.4. Again, there is an excellent agreement between experiment and simulation, and the results also confirm that the RTHS methodology is suitable for the experimental analysis of nonlinear vibrations.

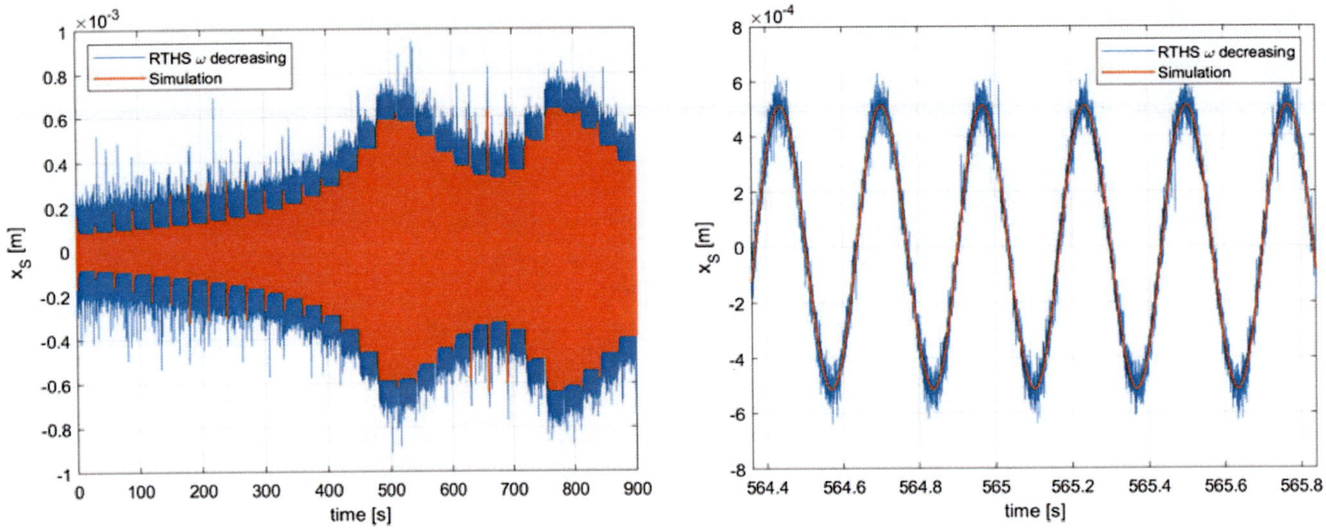

Fig. 13.8 Measured structural displacements of the RTHS absorber testing for harmonic excitation of $a_F = 0.04$ m/s^2 and decreasing excitation frequencies ω: (**a**) entire experiment and (**b**) detail of structural displacement

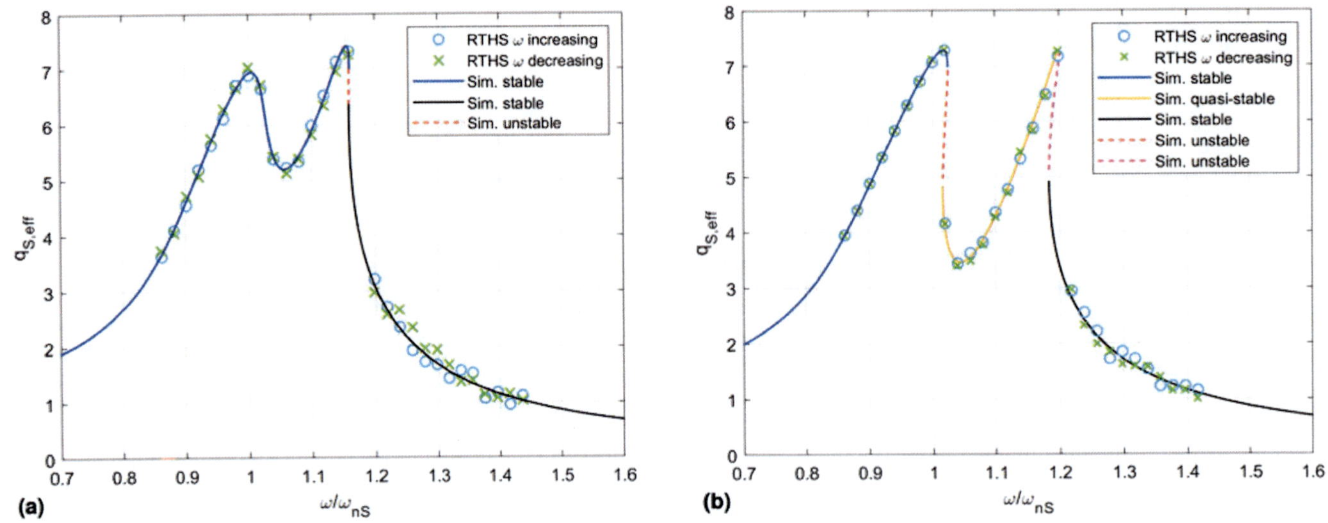

Fig. 13.9 Nonlinear vibration absorption: comparison of FRF amplitudes of numerical simulation and RTHS testing for different excitation levels: (**a**) $a_F = 0.04$ m/s^2 and (**b**) $a_F = 0.08$ m/s^2

13.5 Summary and Conclusion

In this work, real-time hybrid simulation (RTHS) is used to study the vibration mitigation of a nonlinear Duffing-type absorber. Therefore, the absorber's equations of motion are derived and solved, before the concept of hybrid simulation is introduced. In the study, both, structure and dynamic absorber, are assumed to be of Duffing type. Coupling the oscillators renders a set of nonlinear differential equations, which is solved for steady-state excitation using the harmonic balance approximation. For given structural parameter and a chosen mass ratio, the required absorber parameters are determined by the generalized Den Hartog's equal peak method. Contrary to the linear case where analytical design criteria are readily available, the optimization of the nonlinear absorber is done numerically by minimizing the structural FRF. To confirm the simulated results experimentally, a dynamic absorber is built and tested in the laboratory. After validating the required nonlinear behavior, it is used to demonstrate the dynamic absorption in a real-time hybrid simulation configuration. The RTHS methodology enables reliable testing of the physical absorber without having to construct the host structure it is attached to. Instead, all parts not physically available are modelled numerically and coupled to the physical absorber model by proper interfaces. All RTHS experiments show excellent agreement with theoretical results and confirm that RTHS is a reliable technique and of practical importance, when investigating nonlinear dynamic absorption.

References

1. Hochrainer, M.J.: Real-time hybrid testing: challenges and experiences from a teaching point of view. In: Mains, M., Dilworth, B. (eds.) Topics in Modal Analysis & Testing, vol 9, Conference Proceedings of the Society for Experimental Mechanics Series. Springer, New York (2019)
2. Hochrainer, M.J., Puhwein, A.M.: Investigation of nonlinear dynamic phenomena applying real-time hybrid simulation. In: Kerschen, G. (ed.) Nonlinear Structures and Systems, Volume 1, Conference Proceedings of the Society for Experimental Mechanics Series (2019). https://doi.org/10.1007/978-3-030-12391-8_16
3. Saouma, V., Sivaselvan, M.: Hybrid Simulation: Theory, Implementation and Applications. Taylor & Francis Ltd., London (2008)
4. Bursi, O.S., Wagg, D.: Modern testing techniques for structural systems. In: Dynamics and Control CISM International Centre for Mechanical Sciences, vol. 502. Springer, Cham (2008)
5. Insam, C., Göldeli, M., Klotz, T., Rixen, D.J.: Comparison of feedforward control schemes for real-time hybrid substructuring (RTHS). In: Dynamic Substructures, vol. 4, pp. 1–14. Springer, Cham (2021). https://doi.org/10.1007/978-3-030-47630-4_1
6. Insam, C., Kist, A., Rixen, D.J.: High fidelity real-time hybrid substructure testing using iterative learning control. In: ISR 2020; 52th International Symposium on Robotics (2020). ISBN – 978-3-8007-5428-1
7. Wang, Y., Gao, F., Doyle, F.J.: Survey on iterative learning control, repetitive control, and run-to-run control. J. Process Control. **19**, 1589–1600 (2009)
8. Xu, J.-X., Tan, Y.: Linear and Nonlinear Iterative Learning Control. Springer, Berlin (2003)
9. Bristow, D.A., Baron, K.L., Alleyne, A.G.: Iterative learning control. In: Levine, W.S. (ed.) The Control Handbook. CRC Press, New York (2010)
10. Habib, G., Detroux, T., Viguié, R., Kerschen, G.: Nonlinear generalization of Den Hartog's equal-peak method. Mech. Syst. Signal Process. **52–53**, 17–28 (2015). https://doi.org/10.1016/j.ymssp.2014.08.009
11. Den Hartog, J.P.: Mechanical Vibrations. McGraw-Hill, New York (1934)
12. Nayfeh, A.H., Mook, D.T.: Nonlinear Oscillations. Wiley Interscience, New York (1979)
13. Hochrainer, M.J., Puhwein, A.M.: Design and Characterization of a Multi-Purpose Duffing Oscillator with Flexible Parameter Selection, in Nonlinear Structures and Systems. Volume 1, Conference Proceedings of the Society for Experimental Mechanics Series (2020). https://doi.org/10.1007/978-3-030-47709-7_7

Chapter 14
System Equivalent Model Mixing (SEMM): A Modal Domain Formulation

Miha Pogačar, Domen Ocepek, Gregor Čepon, and Miha Boltežar

Abstract A substructuring-based method System Equivalent Model Mixing (SEMM) has recently been introduced as a novel expansion method. Originally it was implemented in the frequency domain and has been proven to have a great potential. The objective of this paper is to introduce M-SEMM, the modal domain formulation of system equivalent model mixing. Its basic formulation is presented, considering either physical or modal constraints between substructures. When modal constraints are applied, the derivation reveals that the resulting M-SEMM formulation is equivalent to System Equivalent Reduction/Expansion Process (SEREP), one of the most established reduction/expansion methods in the modal domain. Further, when considering physical constraints, the resulting formulation can be seen as a potentially useful novel modal expansion method.

Keywords SEMM · SEREP · Model expansion · Hybrid modelling · Dynamic substructuring

14.1 Introduction

Hybrid modeling is becoming increasingly used in the procedure of predicting the structural dynamics of complex products. System Equivalent Model Mixing (SEMM) [1] has recently been introduced in this field as a frequency-based substructuring method. However, an equivalent substructuring process can also be implemented in the modal domain. Although the dynamic substructuring formulations in different domains are theoretically equivalent, the accuracies of the corresponding substructuring predictions tend to differ in real-life applications, especially when experimental models are involved. In this regard, the advantages of modal methods are typically evident when lightly damped structures are considered in the substructuring process. The objective of this paper is to present M-SEMM, system equivalent model mixing formulation in the modal domain.

14.2 SEMM Concept

This section briefly summarizes the concept of System Equivalent Model Mixing (SEMM), originally presented by Klaassen et al. [1]. The substructuring procedure consists of a coupling and a decoupling step between parent, removed, and overlay models (Fig. 14.1):

1. *Parent model*—(par): The dynamics of this model are obtained in a dense set of DoFs, however, reflect less favorable properties.
2. *Removed model*—(rem): A (duplicate) sub-model of the parent model.
3. *Overlay model*—(ov): The dynamics of this model, obtained in a sparse set of DoFs, are a subject to the extension to the dense set of DoFs.
4. *Hybrid Model*: Resulting model of SEMM, which is a sub-model of the substructuring outcome, acquired at the DoFs corresponding to the parent model.

M. Pogačar (✉) · D. Ocepek · G. Čepon · M. Boltežar
University of Ljubljana, Faculty of Mechanical Engineering, Ljubljana, Slovenia
e-mail: miha.pogacar@fs.uni-lj.si

M. Allen et al. (eds.), *Dynamic Substructures, Volume 4*, Conference Proceedings of the Society for Experimental Mechanics Series,
https://doi.org/10.1007/978-3-031-04094-8_14

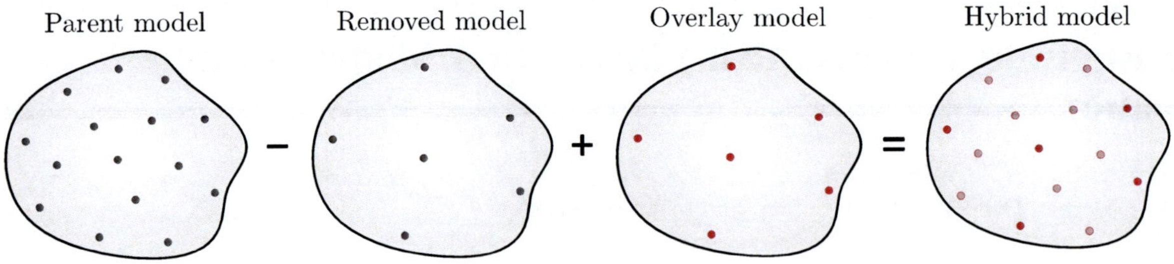

Fig. 14.1 Schematic depiction of SEMM

The dense set of physical DoFs of the parent model can be divided into internal (i) and boundary subset (b), where the boundary subset is collocated with the sparse subset of the removed and overlay model:

$$\boldsymbol{u}^{(\text{par})} = \begin{bmatrix} \boldsymbol{u}_i^{(\text{par})} \\ \boldsymbol{u}_b^{(\text{par})} \end{bmatrix}, \quad \boldsymbol{u}^{(\text{rem})} = \boldsymbol{u}_b^{(\text{rem})}, \quad \boldsymbol{u}^{(\text{ov})} = \boldsymbol{u}_b^{(\text{ov})}.$$

14.3 System Equivalent Model Mixing in Modal Domain

In the following, a set of $m^{(s)}$ mass-normalized free-interface mode shapes $\boldsymbol{\Phi}^{(s)}$ is considered as a reduction matrix for each substructure s. Hence, reduction of physical DoFs $\boldsymbol{u}$ to generalized DoFs $\boldsymbol{\eta}$ formulates as

$$\begin{bmatrix} \boldsymbol{u}_i^{(\text{par})} \\ \boldsymbol{u}_b^{(\text{par})} \\ \boldsymbol{u}_b^{(\text{rem})} \\ \boldsymbol{u}_b^{(\text{ov})} \end{bmatrix} = \begin{bmatrix} \boldsymbol{\Phi}_i^{(\text{par})} & \mathbf{0} & \mathbf{0} \\ \boldsymbol{\Phi}_b^{(\text{par})} & \mathbf{0} & \mathbf{0} \\ \mathbf{0} & \boldsymbol{\Phi}_b^{(\text{rem})} & \mathbf{0} \\ \mathbf{0} & \mathbf{0} & \boldsymbol{\Phi}_b^{(\text{ov})} \end{bmatrix} \begin{bmatrix} \boldsymbol{\eta}^{(\text{par})} \\ \boldsymbol{\eta}^{(\text{rem})} \\ \boldsymbol{\eta}^{(\text{ov})} \end{bmatrix}. \tag{14.1}$$

For the sake of simplicity, damping terms are omitted in further derivations. However, generally, the solution also applies to proportionally damped systems. The un-coupled M-SEMM equation of motion is given as

$$\begin{bmatrix} \mathbf{I}^{(\text{par})} & \mathbf{0} & \mathbf{0} \\ \mathbf{0} & -\mathbf{I}^{(\text{rem})} & \mathbf{0} \\ \mathbf{0} & \mathbf{0} & \mathbf{I}^{(\text{ov})} \end{bmatrix} \begin{bmatrix} \ddot{\boldsymbol{\eta}}^{(\text{par})} \\ \ddot{\boldsymbol{\eta}}^{(\text{rem})} \\ \ddot{\boldsymbol{\eta}}^{(\text{ov})} \end{bmatrix} + \begin{bmatrix} \boldsymbol{\omega}_r^{2\,(\text{par})} & \mathbf{0} & \mathbf{0} \\ \mathbf{0} & -\boldsymbol{\omega}_r^{2\,(\text{rem})} & \mathbf{0} \\ \mathbf{0} & \mathbf{0} & \boldsymbol{\omega}_r^{2\,(\text{ov})} \end{bmatrix} \begin{bmatrix} \boldsymbol{\eta}^{(\text{par})} \\ \boldsymbol{\eta}^{(\text{rem})} \\ \boldsymbol{\eta}^{(\text{ov})} \end{bmatrix} = \begin{bmatrix} \boldsymbol{f}_{\text{m}}^{(\text{par})} \\ \boldsymbol{f}_{\text{m}}^{(\text{rem})} \\ \boldsymbol{f}_{\text{m}}^{(\text{ov})} \end{bmatrix} + \begin{bmatrix} \boldsymbol{g}_{\text{m}}^{(\text{par})} \\ \boldsymbol{g}_{\text{m}}^{(\text{rem})} \\ \boldsymbol{g}_{\text{m}}^{(\text{ov})} \end{bmatrix}, \tag{14.2}$$

where for a substructure s, $\mathbf{I}^{(s)}$ is an identity matrix, and $\boldsymbol{\omega}_r^{2\,(s)}$ is a diagonal matrix of its eigenvalues. In standard SEMM [1], compatibility condition reads as $\boldsymbol{u}_b^{(\text{rem})} = \boldsymbol{u}_b^{(\text{ov})}$ and $\boldsymbol{u}_b^{(\text{rem})} = \boldsymbol{u}_b^{(\text{par})}$ for coupling and decoupling step, respectively. Considering the transformation to the modal domain, **physical compatibility** is approximated as

$$\boldsymbol{\Phi}_b^{(\text{rem})}\,\boldsymbol{\eta}^{(\text{rem})} = \boldsymbol{\Phi}_b^{(\text{ov})}\,\boldsymbol{\eta}^{(\text{ov})}, \tag{14.3}$$

$$\boldsymbol{\Phi}_b^{(\text{rem})}\,\boldsymbol{\eta}^{(\text{rem})} = \boldsymbol{\Phi}_b^{(\text{par})}\,\boldsymbol{\eta}^{(\text{par})}. \tag{14.4}$$

When number of boundary DoFs $n_b \geq m^{(\text{rem})}$, the MCFS approach [2] can also be adopted for weakening the physical constraints, resulting in **modal compatibility**:[1]

$$\mathbf{I}^{(\text{rem})}\,\boldsymbol{\eta}^{(\text{rem})} = \boldsymbol{\Phi}_b^{(\text{rem})\dagger}\,\boldsymbol{\Phi}_b^{(\text{ov})}\,\boldsymbol{\eta}^{(\text{ov})}, \tag{14.5}$$

$$\mathbf{I}^{(\text{rem})}\,\boldsymbol{\eta}^{(\text{rem})} = \mathbf{I}^{(\text{par})}\,\boldsymbol{\eta}^{(\text{par})}. \tag{14.6}$$

[1] Equivalently, one could also consider premultiplying by $\boldsymbol{\Phi}_b^{(\text{rem})\dagger}$. In a typical SEMM application, overlay model is experimentally obtained and we tend to avoid the inversion of experimental data to avoid enhancing the experimental errors.

In either way, compatibility condition can be written as $\mathbf{B}_{\mathrm{m}}\,\boldsymbol{\eta} = \mathbf{0}$, for which the corresponding primal operator can be obtained by a null-space operation $\mathbf{L}_{\mathrm{m}} = \mathrm{null}(\mathbf{B}_{\mathrm{m}})$. Then, the transformation to the unique set of generalized coordinates $\boldsymbol{\xi}$ reads as

$$\boldsymbol{\eta} = \mathbf{L}_{\mathrm{m}}\,\boldsymbol{\xi} = \begin{bmatrix} \mathbf{L}_{\mathrm{m}}^{(\mathrm{par})} \\ \mathbf{L}_{\mathrm{m}}^{(\mathrm{rem})} \\ \mathbf{L}_{\mathrm{m}}^{(\mathrm{ov})} \end{bmatrix} \boldsymbol{\xi}\,. \tag{14.7}$$

The resulting primally assembled M-SEMM mass and stiffness matrix can be obtained as

$$\tilde{\mathbf{M}}_{\mathrm{m}}^{(\mathrm{SEMM})} = \mathbf{L}_{\mathrm{m}}^{\mathrm{T}} \begin{bmatrix} \mathbf{I}^{(\mathrm{par})} & \mathbf{0} & \mathbf{0} \\ \mathbf{0} & -\mathbf{I}^{(\mathrm{rem})} & \mathbf{0} \\ \mathbf{0} & \mathbf{0} & \mathbf{I}^{(\mathrm{ov})} \end{bmatrix} \mathbf{L}_{\mathrm{m}}\,, \tag{14.8}$$

$$\tilde{\mathbf{K}}_{\mathrm{m}}^{(\mathrm{SEMM})} = \mathbf{L}_{\mathrm{m}}^{\mathrm{T}} \begin{bmatrix} \omega_r^{2\,(\mathrm{par})} & \mathbf{0} & \mathbf{0} \\ \mathbf{0} & -\omega_r^{2\,(\mathrm{rem})} & \mathbf{0} \\ \mathbf{0} & \mathbf{0} & \omega_r^{2\,(\mathrm{ov})} \end{bmatrix} \mathbf{L}_{\mathrm{m}}\,, \tag{14.9}$$

where for the (undamped) eigenvalue problem $\left(\tilde{\mathbf{K}}_{\mathrm{m}}^{(\mathrm{SEMM})} - \boldsymbol{\lambda}^{(\mathrm{SEMM})}\,\tilde{\mathbf{M}}_{\mathrm{m}}^{(\mathrm{SEMM})}\right)\boldsymbol{\Phi}_{\xi}^{(\mathrm{SEMM})} = \mathbf{0}$, $\boldsymbol{\lambda}^{(\mathrm{SEMM})}$ are the eigenvalues and $\boldsymbol{\Phi}^{(\mathrm{SEMM})} = \boldsymbol{\Phi}^{(\mathrm{par})}\,\mathbf{L}_{\mathrm{m}}^{(\mathrm{par})}\,\boldsymbol{\Phi}_{\xi}^{(\mathrm{SEMM})}$ are the eigenvectors of the hybrid M-SEMM model.

It is interesting to observe that when modal compatibility is applied, the corresponding primal substructuring operator can be derived as $\mathbf{L}_{\mathrm{m}} = \left[\boldsymbol{\Phi}_b^{(\mathrm{rem})\dagger}\,\boldsymbol{\Phi}_b^{(\mathrm{ov})}\;\; \boldsymbol{\Phi}_b^{(\mathrm{rem})\dagger}\,\boldsymbol{\Phi}_b^{(\mathrm{ov})}\;\; \mathbf{I}^{(\mathrm{ov})}\right]^{\mathrm{T}}$. In this case, the upper eigenvalue problem has a trivial solution, which results in $\boldsymbol{\lambda}^{(\mathrm{SEMM})} = \omega_r^{2\,(\mathrm{ov})}$ and $\boldsymbol{\Phi}^{(\mathrm{SEMM})} = \boldsymbol{\Phi}^{(\mathrm{par})}\,\boldsymbol{\Phi}_b^{(\mathrm{rem})\dagger}\,\boldsymbol{\Phi}_b^{(\mathrm{ov})}$. This reveals that when modal constraints are applied, M-SEMM is analytically equivalent to the well-known and widely established System Equivalent Reduction Expansion Process (SEREP) [3].

On the other hand, the application of physical compatibility requires consideration for two cases. If the number of boundary DoFs $n_b \geq m^{(\mathrm{rem})}$, the excessive number of constraints would result in the reduced number of hybrid model's modes (compared to the overlay model), which typically is of no practical use. The under-determined case ($n_b < m^{(\mathrm{rem})}$) however, represents a potentially useful expansion method in the modal domain and can also be seen as an extension of the established SEREP methodology.

14.4 Numerical Example

To demonstrate the applicability of the proposed method, let us consider an overlay model for which 10 mode shapes were identified from the responses in 10 physical DoFs, as shown in Fig. 14.2a. A numerical parent model (with simplified geometry and boundary conditions) is used to perform the expansion process to the set of 30 DoFs, depicted in Fig. 14.2b.

Assuming a numerical parent model, in principle the expansion can be performed using an arbitrary number of parent mode shapes $m^{(\mathrm{par})}$ (which is equal to $m^{(\mathrm{rem})}$). SEREP is of practical use when $n_b \geq m^{(\mathrm{rem})}$, whereas M-SEMM can be applied for $n_b < m^{(\mathrm{rem})}$. The average modal assurance criterion (MAC) [4] values for 10 expanded overlay modes demonstrate the potential of M-SEMM to achieve high accuracy of expansion (Fig. 14.3).

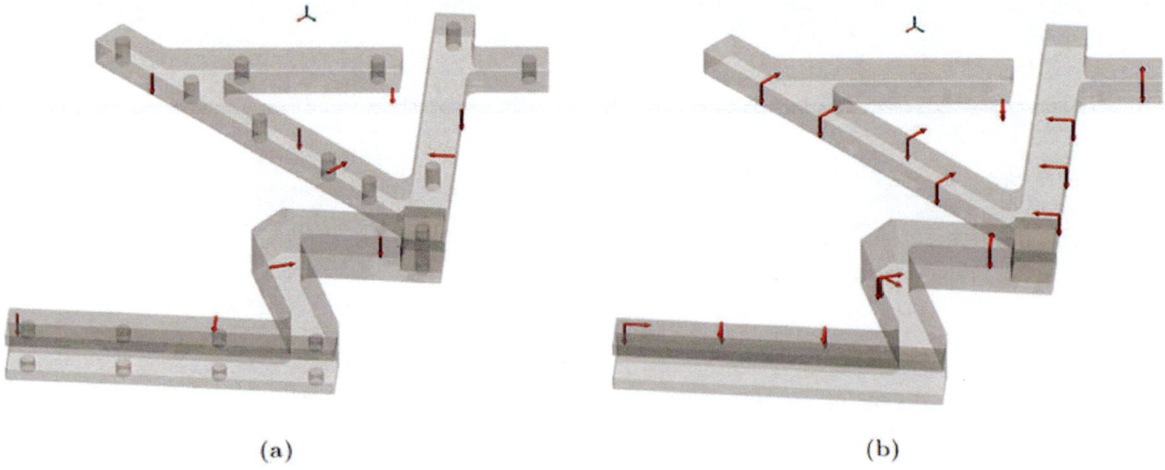

Fig. 14.2 Geometry and physical DoFs: (**a**) Overlay model; (**b**) Parent model

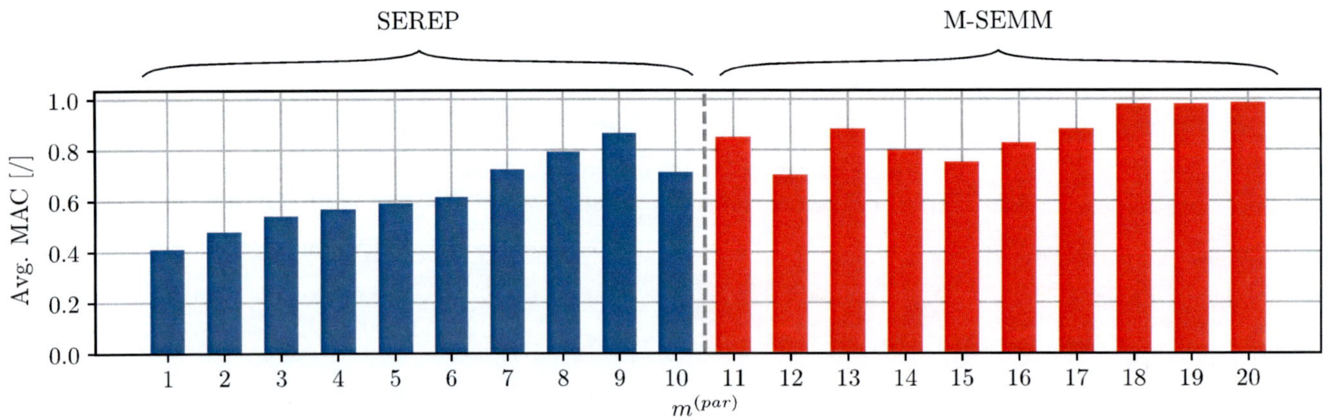

Fig. 14.3 Elementary example—depiction of SEMM models

References

1. Klaassen, S.W., van der Seijs, M.V., de Klerk, D.: System equivalent model mixing. Mech. Syst. Signal Process. **105**, 90–112 (2018)
2. Allen, M.S., Mayes, R.L., Bergman, E.J.: Experimental modal substructuring to couple and uncouple substructures with flexible fixtures and multi-point connections. J. Sound Vib. **329**(23), 4891–4906 (2010)
3. O'Callahan, J.C., Avitabile, P., Riemer, R.: System equivalent reduction expansion process (SEREP). In: Proceedings of the VII International Modal Analysis Conference (IMAC), vol. 1, pp. 29–37. Society for Experimental Mechanics, Bethel (1989)
4. Allemang, R.J.: The modal assurance criterion–twenty years of use and abuse. Sound Vibration **37**(8), 14–23 (2003)

Chapter 15
Hybrid Testing of a Cantilever Beam with Two Controlled Degrees of Freedom

Alessandra Vizzaccaro, Sandor Beregi, David Barton, and Simon Neild

Abstract Hybrid testing of a cantilever beam with two controlled degrees of freedom at the interface between virtual and physical structures is performed in this work. To circumvent instabilities caused by delays in the transfer system and inaccuracies caused by error in the characterization of the transfer system, an alternative open-loop method based on harmonic balance method is adopted that is suitable for steady-state motion. The periodic solution that ensures synchronous motion of the interface between the virtual and the physical structure is found via quasi-Newton iterations. Results show a strong agreement between the two sides of the interface, thus proving the capabilities of the method in dealing with continuous structures with two controlled degrees of freedom.

Keywords Hybrid testing · Harmonic balance method · Substructuring · Wing-like structures

15.1 Introduction

Hybrid testing is an experimental testing method that couples a virtual substructure with a physical one in real time. The two are coupled through a controlled interface that must ensure compatibility and equilibrium of forces and displacements, thereby reproducing the behavior of the true assembly. This allows a small, critical part of a structure to be tested physically within the context of the whole structure. Furthermore, such physical substructure could be tested in different virtual environments, thus helping the design of engineering structures. The main motivation of this work is to assess the feasibility of hybrid testing of wing-like structures in wind tunnels, where the portion of the wing containing the root, easier to model numerically, is the virtual structure and that containing the wing tip is the physical one. From a control point of view, the main challenge of such a test is dealing with a continuous structure with more than one degree of freedom to control.

15.2 Test Rig

A prototype test rig has been built with the aim of enabling future testing of wings. The structure under investigation is a stainless steel cantilever beam with rectangular cross section (1 mm × 25.4 mm), clamped on the virtual side and free on the physical side. The position of the interface can be varied by changing the length of the virtual substructure, whereas the length of the physical one is fixed to 360 mm. In Fig. 15.1, a sketch of the test rig is shown. The rig consists of two shakers actuating a clamp that controls the two degrees of freedom, the vertical and angular displacement, of the cantilever interface. The remaining degrees of freedom are constrained by a vertical slider coupled with a rotational joint coaxial with the interface section. Two force transducers placed on the clamp and two lasers measuring the clamp displacement allow deriving force, moment, displacement, and rotation of the interface.

A. Vizzaccaro (✉) · S. Beregi · D. Barton
Faculty of Engineering, School of Computer Science, Electrical and Electronic Engineering, and Engineering Mathematics, University of Bristol, Bristol, UK
e-mail: yo20506@bristol.ac.uk; alessandra.vizzaccaro@bristol.ac.uk; sandor.beregi@bristol.ac.uk; david.barton@bristol.ac.uk

S. Neild
Faculty of Engineering, School of Civil, Aerospace and Mechanical Engineering, University of Bristol, Bristol, UK
e-mail: simon.neild@bristol.ac.uk

M. Allen et al. (eds.), *Dynamic Substructures, Volume 4*, Conference Proceedings of the Society for Experimental Mechanics Series,
https://doi.org/10.1007/978-3-031-04094-8_15

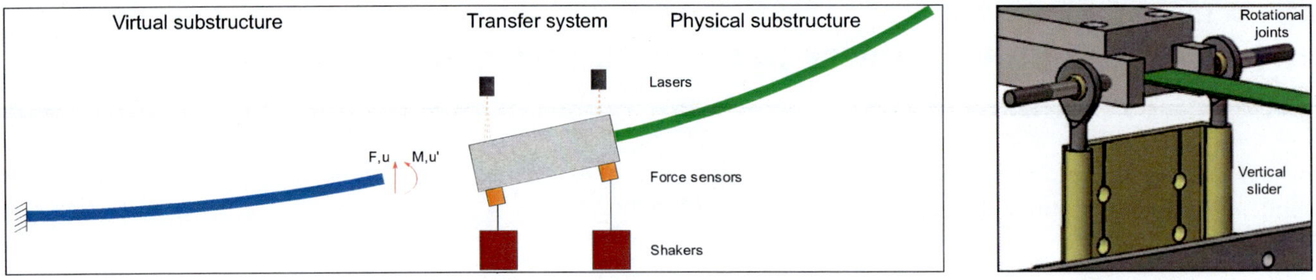

Fig. 15.1 Schematic overview of the hybrid experiment (left) and detail of the slider-joint constraint at the physical interface (right)

15.3 Background

In conventional real-time hybrid testing, the motion of the interface is controlled in real time, and equilibrium and compatibility conditions between the virtual and physical sides of the assembly are enforced, generating a closed-loop system that involves not only the virtual and physical structures but also the transfer system. Time delays, always present in the transfer system, can produce dynamic instability of the closed-loop system [1–4]. Moreover, an inaccurate characterization of the transfer system would invariably result in discrepancies between the behavior of the interface at the virtual side and that at the physical side [1–4]. Preliminary stability analysis on a numerical model of the present setup has shown that for nearly every partition of the beam, even for small values of the time delay in the transfer system, the closed-loop system is unstable. Moreover, translational and rotational joints in the test rig introduce nonlinearities due to friction in the transfer system, which thus becomes more difficult to characterize and compensate.

15.4 Method

To overcome the abovementioned stability and accuracy issues, an alternative open-loop method is proposed. The method assumes a periodic regime of vibration of the whole assembly when subjected to a periodic excitation. Based on such assumption, harmonic balance method is applied to the hybrid experiment to impose convergence between the behavior of the interface at the virtual side and that at the physical side. Quasi-Newton methods are applied to find the zero of the harmonic residual $R(V)$ which is a function of the harmonic coefficients of the input voltage to the shakers V that reads:

$$\boldsymbol{R}\,(\boldsymbol{V}) = \boldsymbol{Z}_N\,\boldsymbol{U}_m\,(\boldsymbol{V}) - \boldsymbol{F}_m\,(\boldsymbol{V}) - \boldsymbol{F}_{\mathrm{EXT}}$$

where Z_N is the dynamic stiffness matrix representing the response of the numerical structure condensed at the interface section, F_{EXT} are the harmonic coefficients of the excitation force, and $U_m(V)$ and $F_m(V)$ are the harmonic coefficients of the measured displacements and forces at the interface which depend on the input voltage V. One major advantage of the method is that it does not require any knowledge of the transfer system, in that nonlinear forces are automatically balanced by the multi-harmonic motion of the shakers. Errors introduced by inaccuracies in the characterization of the transfer system are thus eliminated, leaving the only source of inaccuracies to be the harmonic truncation and the presence of noise in the measurements.

15.5 Results

Results obtained with HBM up to 3 harmonics on a 500-mm-long hybrid cantilever beam (360 mm in the physical side, 140 mm in the virtual side) excited with a vertical sinusoidal point load at 4.5 Hz on the virtual side are reported in Fig. 15.2. The harmonic coefficients of the interface displacement and rotation of the virtual side subjected to the measured and external forces are compared with those from the measured interface displacement and rotation. A very good agreement is shown, with some discrepancies appearing on the higher-order harmonics of the measured force and moment. Convergence to the desired tolerance (1% of the initial residual) is attained within a few iterations, thus making the present approach

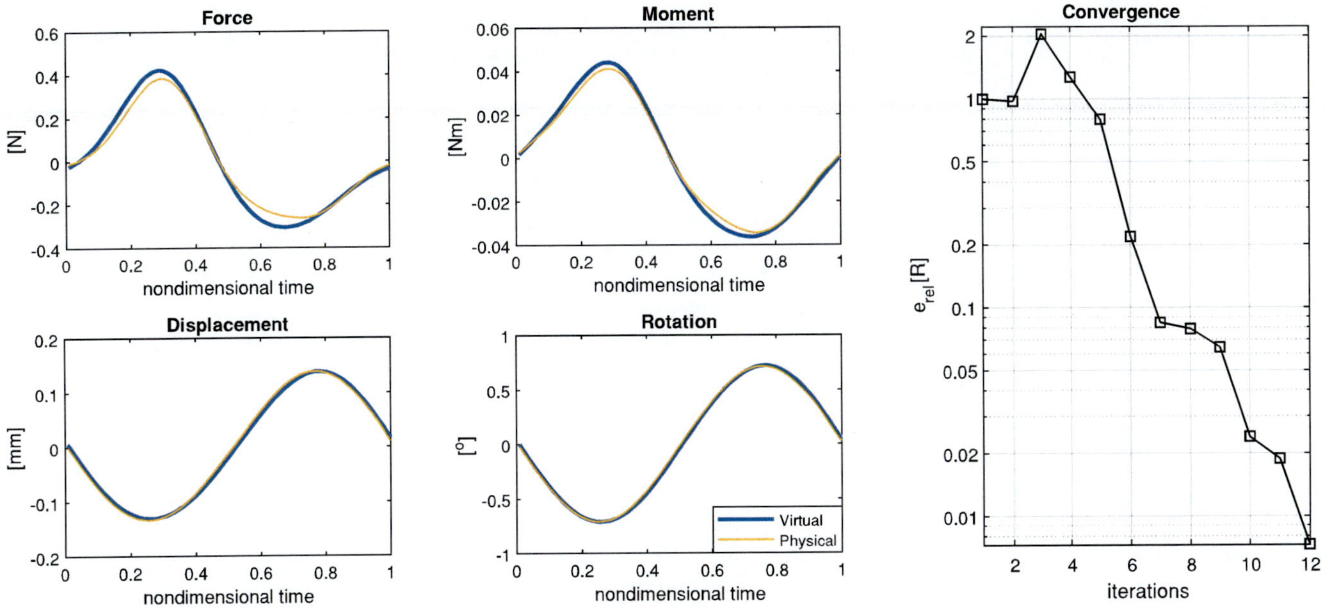

Fig. 15.2 Comparison between virtual and physical quantities at the interface (left). Rate of convergence (right)

an attractive alternative to closed-loop controlled real-time hybrid testing for the case where steady-state dynamics are of interest.

15.6 Conclusion

In this work, hybrid testing is performed on a cantilever beam with two controlled degrees of freedom: displacement and rotation. A test rig has been designed and built to demonstrate the feasibility of hybrid testing of wing-like structures, which serves as a proof of concept for future testing of wings, where only the portion containing the wing tip is experimentally tested. The potential instabilities arising in classical real-time hybrid testing of cantilever structures are avoided by adopting an open-loop iterative method based on the assumption of periodic steady-state response. Results show that the method can deal with the challenge of controlling two actuators, both in multi-harmonic response, in a robust manner, and with fast rate of convergence.

Acknowledgments This work is funded by the *Digital twins for improved dynamic design* (EP/R006768/1) EPSRC grant. The authors are thankful for the support.

References

1. Blakeborough, A., Williams, M.S., Darby, A.P., Williams, D.M.: The development of real-time substructure testing. Phil. Trans. R. Soc. A. **359**, 1869–1891 (2001)
2. Mosqueda, G., Stojadinović, B., Mahin, S.A.: Real-time error monitoring for hybrid simulation. Part I: methodology and experimental verification. J. Struct. Eng. **133**(8), 1100–1108 (2007)
3. Terkovics, N., Neild, S.A., Lowenberg, M., Szalai, R., Krauskopf, B.: Substructurability: the effect of interface location on a real-time dynamic substructuring test. Proc. R. Soc. A. **472**, 20160433 (2016)
4. Insam, C., Rixen, D.J.: Fidelity assessment of real-time hybrid substructure testing: a review and the application of artificial neural networks. Exp. Tech. **46**, 137–152 (2021)

Chapter 16
Experimental Substructuring of the Dynamic Substructures Round-Robin Testbed

D. Roettgen, G. Lopp, A. Jaramillo, and B. Moldenhauer

Abstract Experimental-analytical substructuring has been a popular field of research for several years and has seen many great advances for both frequency-based substructuring (FBS) and component mode synthesis (CMS) techniques. To examine these technical advances, a new benchmark structure has been designed through the SEM dynamic substructuring technical division to act as a benchmark study for anyone researching in the field. This work contains the first attempts at experimental dynamic substructuring using the new SEM testbed. Complete dynamic substructuring predictions will be presented along with an assessment of variability and nonlinear response in the testbed assembly. Systems will be available to check out through the authors beginning in December of 2021, and this paper intends to initiate in full the round-robin challenge.

Keywords Dynamic substructuring · Experimental analytical substructuring · Component mode synthesis · Frequency-based substructuring · Structural modification

16.1 Benchmark Structure

Tremendous progress has been made with research surrounding experimental-analytical substructuring when the SEM community previously rallied around a common testbed. In 2011, the experimental substructures focus group selected a testbed structure, the Ampair 600 Wind Turbine. The origin of this testbed is discussed in detail in [1]. Many researchers have studied this benchmark structure; see [2–9]. During these studies, several new methods and technologies were developed, but the Ampair 600 Wind Turbine turned out to be quite a challenging structure due to the complexity of joints and the nature of utilizing a structure not designed within the community. Despite these challenges, the community made great strides in the field of dynamic substructuring.

At IMAC-XXXVI, the focus group on Dynamic Substructuring officially transitioned into a Technical Division of SEM titled Dynamic Substructures. During this inaugural meeting, the Technical Division discussed the creation of a new benchmark structure to kick off a new era of research in experimental analytical substructuring. It was determined that a new benchmark structure would be an appropriate test of the current capabilities of the Dynamic Substructures Technical Division and would foster future collaboration. A group of researchers collaborated to generate the design of a new round-robin test vessel for dynamic substructuring. After a few design iterations, the four-unit frame (shown in Fig. 16.1) was selected in an airplane-like configuration (shown in Fig. 16.2).

Since finalizing the design, eight frames and eight sets of wings (for each thickness) have been manufactured. Measurements, modes, weights, and detailed assembly instructions from these frames will be available on the Dynamic

Sandia National Laboratories is a multimission laboratory managed and operated by National Technology and Engineering Solutions of Sandia, LLC., a wholly owned subsidiary of Honeywell International, Inc., for the U.S. Department of Energy's National Nuclear Security Administration under contract DE-NA-0003525. This paper describes objective technical results and analysis. Any subjective views or opinions that might be expressed in the paper do not necessarily represent the views of the U.S. Department of Energy or the United States Government.

D. Roettgen (✉) · G. Lopp · A. Jaramillo
Sandia National Laboratories, Albuquerque, NM, USA
e-mail: drroett@sandia.gov; garret.lopp@sandia.gov; antjara@sandia.gov

B. Moldenhauer
University of Wisconsin-Madison, Madison, WI, USA
e-mail: bmoldenhauer@wisc.edu

M. Allen et al. (eds.), *Dynamic Substructures, Volume 4*, Conference Proceedings of the Society for Experimental Mechanics Series,
https://doi.org/10.1007/978-3-031-04094-8_16

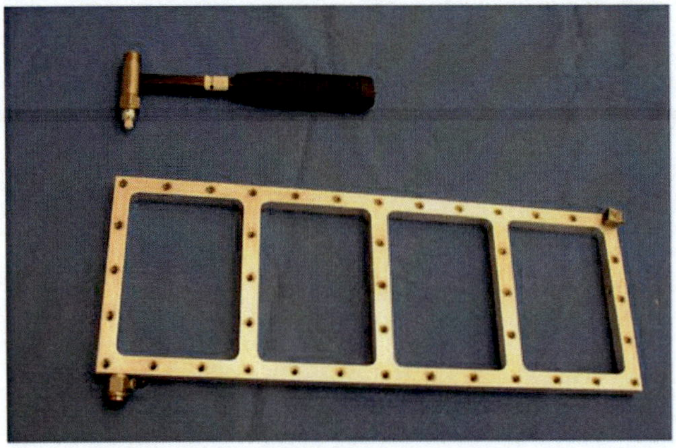

Fig. 16.1 Four unit frame

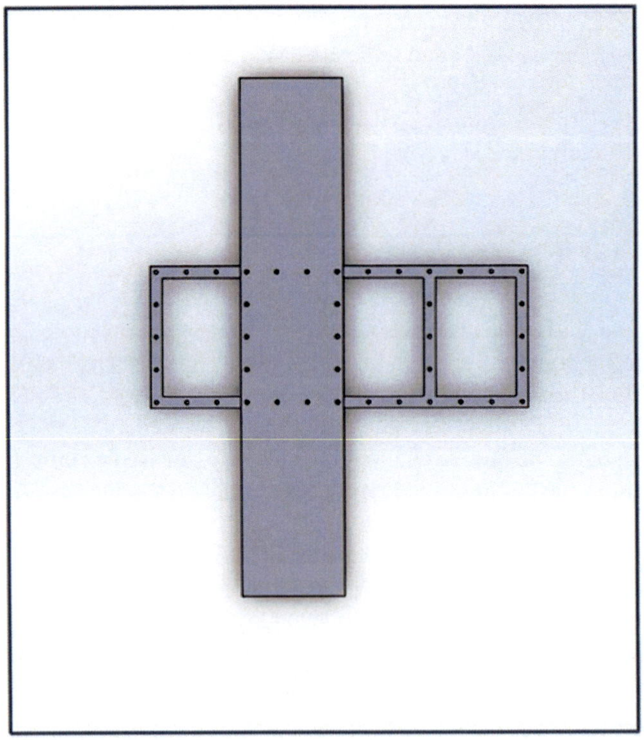

Fig. 16.2 Airplane configuration

Substructuring Wiki. To ensure that this structure is suitable for dynamic substructuring, the team at Sandia National Labs has completed an experimental substructuring prediction with the hardware.

16.2 Example

The goal of this substructuring exercise was to ensure that substructuring of the new round-robin structure was possible before sending out the set of units to interested researchers. Multiple substructuring configurations have been tested, and one is shown here for brevity. The intent of this exercise is to connect an experimental model of the thin wing to an experimental model of the frame. The hardware used included frame SN001, thin wing SN WING006A, and a plate acting as a transmission simulator. This hardware was assembled using four steel fasteners, one at each corner, with washers between

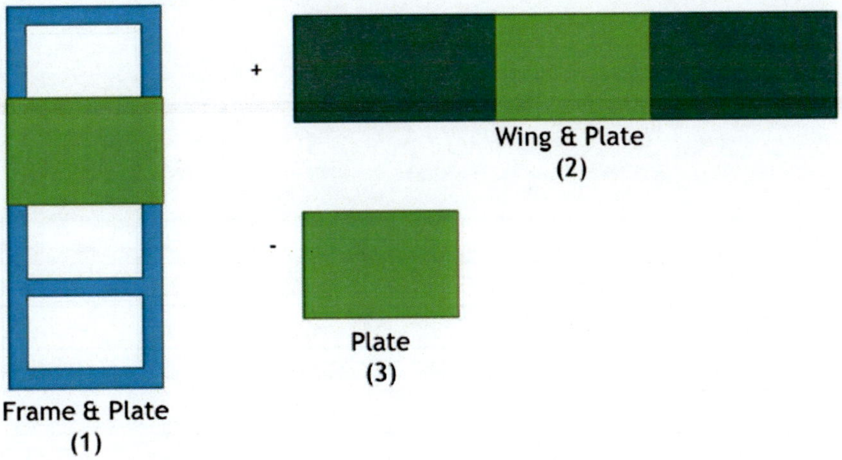

Fig. 16.3 Substructure configurations

the subcomponents. Experiments were performed on substructures 1, 2, and 3 shown in Fig. 16.3 using a laser Doppler vibrometer.

Transmission simulator theory [10] was used where a fixture plate is attached near the connecting hardware. Note that a transmission simulator was attached to both the frame and wing in this example, but only one fixture plate was removed – thus, the truth comparison also includes one fixture plate. A summary of the substructuring technique is repeated here for reference. First, measurements were taken on all three systems, and a set of uncoupled block-diagonal equations of motions was written using the measured modal parameters.

$$\begin{bmatrix} I_1 & 0 & 0 \\ 0 & I_2 & 0 \\ 0 & 0 & -I_3 \end{bmatrix} \begin{Bmatrix} \ddot{q}_1 \\ \ddot{q}_2 \\ \ddot{q}_3 \end{Bmatrix} + \begin{bmatrix} \left[\ddots \omega_1^2 \ddots\right] & 0 & 0 \\ 0 & \left[\ddots \omega_2^2 \ddots\right] & 0 \\ 0 & 0 & -\left[\ddots \omega_3^2 \ddots\right] \end{bmatrix} \begin{Bmatrix} q_1 \\ q_2 \\ q_3 \end{Bmatrix} = \begin{Bmatrix} \phi_1^T F_1 \\ \phi_2^T F_2 \\ \phi_3^T F_3 \end{Bmatrix} \tag{16.1}$$

A set of constraints is defined linking motion on the transmission simulator plate in all three structures.

$$\begin{bmatrix} \phi_3^+ & 0 \\ 0 & \phi_3^+ \end{bmatrix} \begin{bmatrix} \phi_1 & 0 & -\phi_3 \\ 0 & \phi_2 & -\phi_3 \end{bmatrix} \begin{Bmatrix} q_A \\ q_B \\ q_{TS} \end{Bmatrix} = \tilde{B} \begin{Bmatrix} q_1 \\ q_2 \\ q_3 \end{Bmatrix} = 0 \tag{16.2}$$

A synthetization matrix L is found such that the constraints are non-arbitrary.

$$L = \text{null}\left(\tilde{B}\right) \tag{16.3}$$

This matrix is used to transform the original equations of motion into a coupled prediction.

Table 16.1 Substructuring prediction results

Elastic Mode Index	Truth		Prediction		Error	
	fn	zt	fn	zt	fn	zt
1	61.91	0.231%	55.76	0.033%	-9.93%	-85.83%
2	107.19	0.185%	96.16	0.123%	-10.29%	-33.70%
3	230.00	0.073%	225.92	0.069%	-1.77%	-5.42%
4	277.66	0.143%	272.85	0.045%	-1.73%	-68.61%
5	341.56	0.246%	330.92	0.104%	-3.12%	-57.59%
6	369.06	0.131%	352.90	0.072%	-4.38%	-45.50%
7	418.91	0.145%	396.28	0.067%	-5.40%	-53.61%
8	499.06	0.251%	598.74	0.100%	19.97%	-59.96%
9	656.09	0.103%	611.07	0.116%	-6.86%	12.57%
10	696.41	0.095%	688.10	0.164%	-1.19%	72.45%
11	798.75	0.029%	726.60	0.093%	-9.03%	215.58%
12	807.50	0.084%	733.88	0.069%	-9.12%	-17.92%
13	857.19	0.197%	784.65	0.119%	-8.46%	-39.35%
14	976.25	0.182%	784.65	0.119%	-19.63%	-34.25%

$$L^T \begin{bmatrix} I_1 & 0 & 0 \\ 0 & I_2 & 0 \\ 0 & 0 & -I_3 \end{bmatrix} L \begin{Bmatrix} \ddot{q}_1 \\ \ddot{q}_2 \\ \ddot{q}_3 \end{Bmatrix} + L^T \begin{bmatrix} \begin{bmatrix} \ddots & & \\ & \omega_1^2 & \\ & & \ddots \end{bmatrix} & 0 & 0 \\ 0 & \begin{bmatrix} \ddots & & \\ & \omega_2^2 & \\ & & \ddots \end{bmatrix} & 0 \\ 0 & 0 & -\begin{bmatrix} \ddots & & \\ & \omega_3^2 & \\ & & \ddots \end{bmatrix} \end{bmatrix} L \begin{Bmatrix} q_1 \\ q_2 \\ q_3 \end{Bmatrix} = L^T \begin{Bmatrix} \phi_1^T F_1 \\ \phi_2^T F_2 \\ \phi_3^T F_3 \end{Bmatrix} \tag{16.4}$$

Modes were retained to 1000 Hz for the substructures 1 and 2, while only rigid modes were retained for substructure 3. Table 16.1 contains the blind substructuring predictions and their accuracy compared to a truth test, and Fig. 16.4 shows a comparison from measured truth to predicted shapes through a MAC plot. Our quality metric was set such that frequency errors under 10% were considered quality, while damping errors under 50% were considered quality. Most frequency errors were in the quality range with two modes outside in the first 700 Hz; the damping quality was slightly poorer as expected and may be corrected by fine-tuning subcomponent damping. Additional substructuring configurations and predictions were completed and will be presented as a part of the IMAC presentation.

16.3 Kick-Off

The Dynamic Substructures Technical Division Round-Robin is now ready to begin. Please contact Dan Roettgen at to sign up. For detailed instructions on the challenges, please visit the Dynamic Substructuring Wiki during and after IMAC-XL. The results presented in this work and presentation capture a small fraction of the research that will be completed using this new benchmark structure. We look forward to participation and collaborations among research groups for years to come.

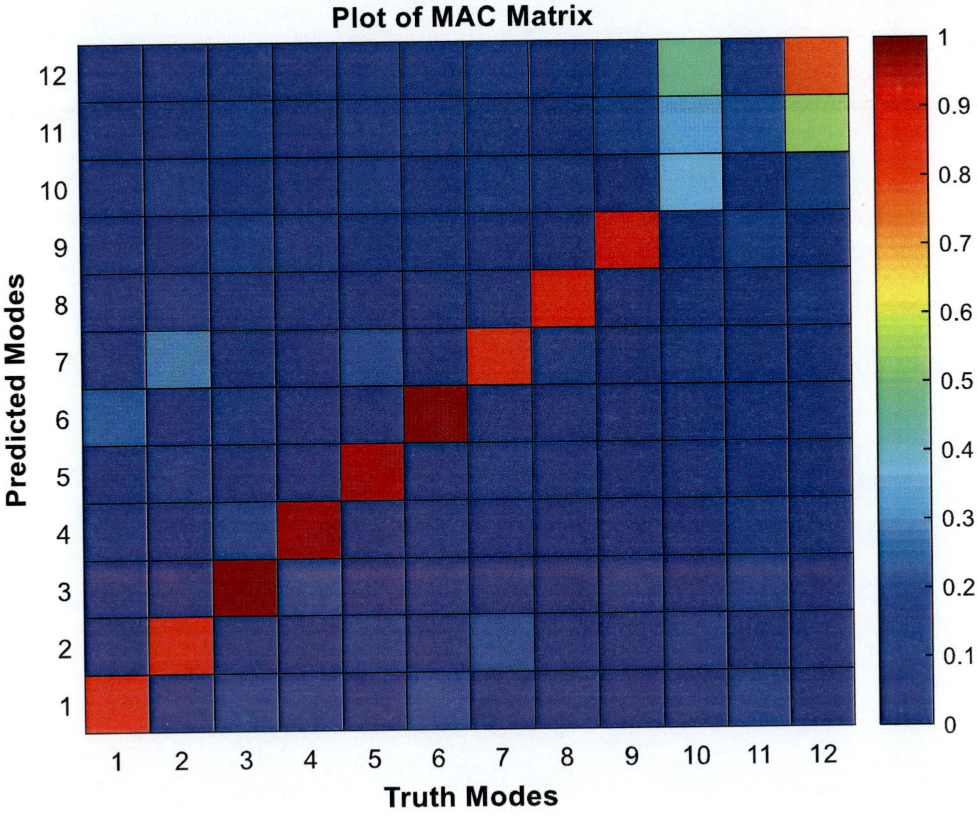

Fig. 16.4 MAC of prediction results and truth results

References

1. Mayes, R.L.: An introduction to the SEM substructures focus group test Bed – The Ampair 600 wind turbine. In: Topics in Experimental Dynamics Substructuring & Wind Turbine Dynamics, Jacksonville, FL (2012)
2. Rohe, D.P., Mayes, R.L.: Coupling of a bladed hub to the tower of the ampair 600 wind turbine using the transmission simulator method. In: Topics in Experimental Dynamics Substructuring, Vol. 2, Orlando, FL (2014)
3. Roettgen, D.R., Mayes, R.L.: Ampair 600 wind turbine 3-bladed assembly substructuring using the transmission simulator method. In: Proceedings from the XXXIII IMAC Conference, Orlando, FL (2015)
4. Rahimi, S., deKlerk, D., Rixen, D.J.: The ampair 600 wind turbine benchmark: results from the frequency based substructuring applied to the rotor assembly. In: Proceedings of the 31st International Modal Analysis Conference, Orlando, FL (2013)
5. Brunetti, J., Culla, A., D'Ambrogio, W., Fregolent, A.: Experimental dynamic substructuring of the ampair wind turbine test bed. In: Proceedings of the International Modal Analysis Conference, Orlando, FL (2014)
6. Allen, M.S., Gindlin, H.M., Mayes, R.L.: Experimental modal substructuring to estimate fixed-base modes from a test on a flexible fixture. Journal of Sound and Vibration. **330**(18-19), 4413–4428 (2011)
7. Harvie, J., Avitable, P.: Comparison of some wind turbine blade tests in various configurations. In: Proceedings from the 30th International Modal Analysis Conference, Orlando, FL (2012)
8. Gross, J., Seeger, B., Peter, S., Reuss, P.: Applying the transmission simulator techniques to the ampair 600 wind turbine testbed. In: Dynamics of Coupled Structures, Volume 4: Proceedings of the 34th IMAC, Orlando, FL (2016)
9. Cwenarkiewicz, M., Johansson, T.: Experimental dynamic substructuring of an ampair 600 wind turbine hub together with two blades. Linnaeus University Thesis Under Advisor: Andreas Linderholt (2016)
10. Allen, M., Mayes, R., Bergman, E.: Experimental modal substructuring to couple and uncouple substructures with flexible fixtures and multi-point connections. Journal of Sound and Vibration. **329**, 4891–4906 (2010)

Chapter 17
Feasibility of Configuration-Dependent Substructure Decoupling

Jacopo Brunetti, Walter D'Ambrogio, and Annalisa Fregolent

Abstract Substructure decoupling allows identifying the unknown dynamic behavior of a subsystem starting from the dynamic behavior of the whole system and that of the residual part of the system. Recently, the coupling problem has been extended to deal with time or configuration-dependent coupling conditions. This approach is useful to numerically investigate the dynamics of a configuration-dependent system with a reduction of the computational burden. In this paper, we want to examine the feasibility of performing substructure decoupling when the coupling conditions among invariant mechanical subsystems are configuration-dependent. Typical examples of such systems could be a lifting crane or a Cartesian robot. Taking for granted that the dynamic behavior of the whole system must be known for each configuration, several further questions have to be addressed: if it is necessary to take FRF measurements on the connecting DoFs; if, for the residual system, it is necessary to consider a different set of FRF measurements for each configuration; if it is possible to take advantage of the redundancy of information provided by considering multiple configurations, i.e. multiple internal constraints. In the paper, we will try to answer these questions by starting from a well-known test bed, previously used for substructure decoupling.

Keywords Experimental dynamic substructuring · Substructure decoupling · Configuration Dependent substructuring

17.1 Introduction

Dynamic substructuring aims to find the dynamic behavior of coupled systems starting from the dynamic behavior of the component subsystems (coupling problem) and, conversely, to find the unknown dynamic behavior of a given subsystem starting from the known dynamic behavior of the assembled system and of the residual subsystem (decoupling problem). A general framework for dynamic substructuring is provided in [1]; the coupling problem is tackled in [2–4]; the decoupling problem is tackled in [5–9].

Although the classical substructuring approach assumes that the system is time invariant, time-variant or configuration dependent systems composed of invariant mechanical subsystems subjected to time or configuration dependent coupling conditions have been tackled with this approach [10–13]. Results in [14–16] highlight that numerical analysis of configuration dependent problems in the framework of dynamic substructuring provides meaningful information reducing the computational effort.

Starting from this framework, in this paper the decoupling problem of invariant mechanical subsystems with configuration dependent coupling conditions is considered. Typical examples of such systems could be a lifting crane or a Cartesian robot. The dynamic behavior of the whole system must be known for each configuration, but it is assumed that the component subsystems are invariant. Therefore, the use of multiple configurations can provide a more rich set of information that can improve the identification of the unknown (invariant) subsystem. Moreover, in substructure decoupling it is not required to know the dynamic behavior of the assembled system at the connecting DoFs between the component subsystems [9, 17]. Several questions are addressed in the present paper: if it is necessary to take FRF measurements on some connecting DoF;

J. Brunetti (✉) · W. D'Ambrogio
Dipartimento di Ingegneria Industriale e dell'Informazione e di Economia, Università dell'Aquila, L'Aquila, Italy
e-mail: jacopo.brunetti@univaq.it; walter.dambrogio@univaq.it

A. Fregolent
Dipartimento di Ingegneria Meccanica e Aerospaziale, Università di Roma La Sapienza, Rome, Italy
e-mail: annalisa.fregolent@uniroma1.it

M. Allen et al. (eds.), *Dynamic Substructures, Volume 4*, Conference Proceedings of the Society for Experimental Mechanics Series,
https://doi.org/10.1007/978-3-031-04094-8_17

if, for the residual system, it is necessary to consider a different set of FRF measurements for each configuration; if it is possible to take advantage of the redundancy of information provided by considering multiple configurations, i.e. multiple internal constraints.

In this paper, decoupling in the frequency domain is applied to obtain the dynamics of the same unknown substructure in three different configurations of the assembled structure that derives from a well-known test bed [9, 17].

17.2 Substructure Decoupling

Substructure decoupling allows identifying the unknown dynamic behavior of a subsystem, starting from the dynamic behavior (e.g. FRFs) of the assembled system RU and that of a known portion of it, the so-called residual subsystem R). The unknown substructure U (N_U DoFs) is joined to the residual substructure R (N_R DoFs) by n_c coupling DoFs through which constraint forces (and moments) are exchanged (see Fig. 17.1). The degrees of freedom of the assembled structure (N_{RU} DoFs) can be partitioned into coupling DoFs (c), internal DoFs of substructure U (u), and internal DoFs of substructure R (r). The FRFs of the unknown subsystem U can be predicted from those of the assembled system RU by taking out the dynamic effect of the residual subsystem R. In principle, this can be accomplished by considering a negative structure, i.e. by adding to the assembled system RU a fictitious subsystem with a dynamic stiffness opposite to that of the residual subsystem R (Fig. 17.2). The effect of the negative system is to add disconnection forces (and moments) to the external forces acting on the assembled system in order to uncouple the unknown subsystem from the assembled system.

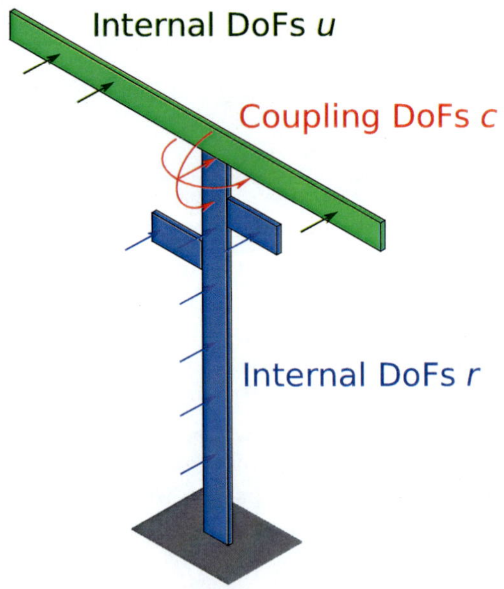

Fig. 17.1 Assembled system RU, with the unknown subsystem U (green) and the residual subsystem R (blue)

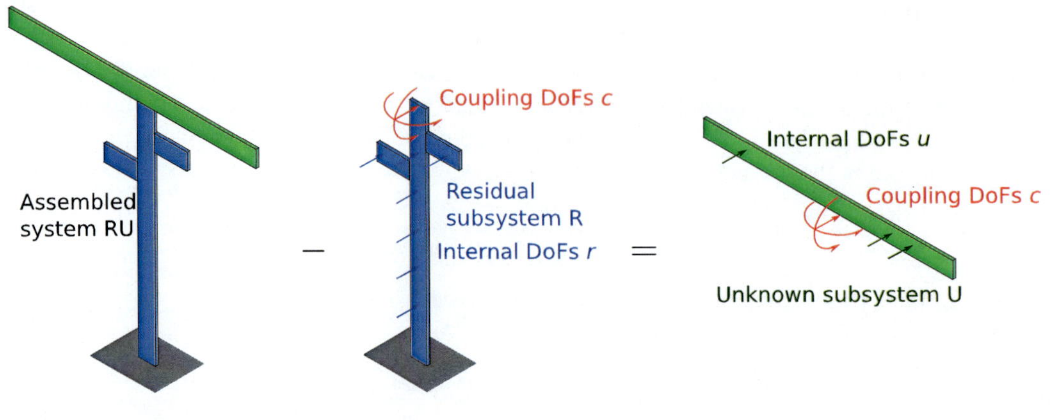

Fig. 17.2 Scheme of the direct decoupling problem

Therefore, the dynamic equilibrium of the assembled system RU is expressed as

$$\left[Z^{RU}\right]\left\{u^{RU}\right\} = \left\{f^{RU}\right\} + \left\{g^{RU}\right\} \tag{17.1}$$

where $\{g^{RU}\}$ is the vector of disconnection forces applied to the assembled system by the negative subsystem, $[Z^{RU}]$ is the dynamic stiffness matrix of the assembled system RU, $\{u^{RU}\}$ is the vector of degrees of freedom of the assembled system RU, and $\{f^{RU}\}$ is the external force vector on the assembled system RU.

In a similar way, the dynamic equilibrium of the negative subsystem is expressed as:

$$-\left[Z^{R}\right]\left\{u^{R}\right\} = \left\{f^{R}\right\} + \left\{g^{R}\right\} \tag{17.2}$$

where $-[Z^R]$ is the dynamic stiffness matrix of the negative subsystem, and $\{u^R\}$, $\{f^R\}$, $\{g^R\}$ are defined as for the assembled system.

17.2.1 Possible Sets of Interface DoFs

It must be noted that the set of disconnection forces is not unique. In fact, several sets of disconnection forces can be devised:

- A trivial set, consisting of disconnection forces acting at the coupling DoFs and opposite to the constraint forces (see Fig. 17.3); in this case, disconnection forces may include moments opposite to the constraint moments.
- Non-trivial sets of disconnection forces acting at different DoFs but able to cancel the constraint forces at the coupling DoFs (see Fig. 17.4); in this case, disconnection forces applied to internal DoFs must be able to provide a moment about the rotation axes.

The set on which the disconnecting forces are acting is the set of interface DoFs. In order that Eqs. (17.1)–(17.2) can be put together to represent the unknown subsystem U, disconnection forces $\{g^{RU}\}$ and $\{g^R\}$ must be in equilibrium, and compatibility between degrees of freedom $\{u^{RU}\}$ and $\{u^R\}$ must hold at the interface between the assembled system RU and the negative subsystem. Therefore, several options for interface DoFs can be considered:

- Standard interface, including only the coupling DoFs (c) between subsystems U and R, e.g. those corresponding to the disconnection forces (and moments) in Fig. 17.3.
- Extended interface, including also a subset of internal DoFs ($i \subseteq r$) of subsystem R.
- Mixed interface, including subsets of coupling DoFs ($d \subset c$) and internal DoFs ($i \subseteq r$), e.g. those corresponding to the disconnection forces in Fig. 17.4 (left).

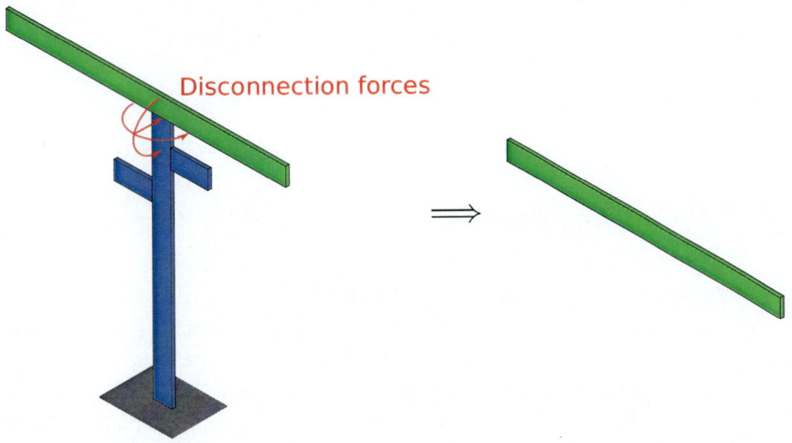

Fig. 17.3 Trivial set of disconnection forces (and moments) acting on the assembled structure (corresponding to the standard interface)

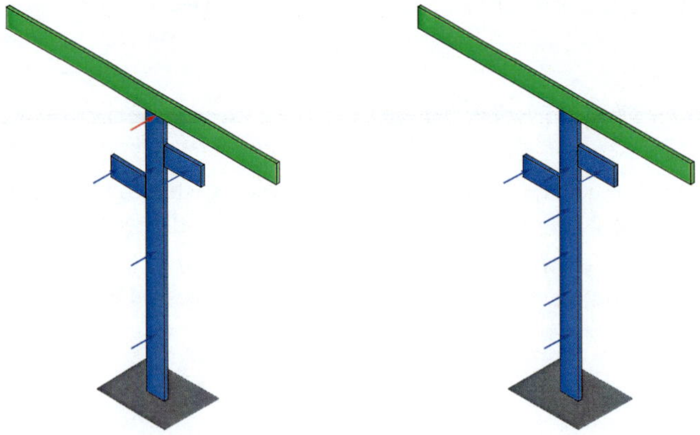

Fig. 17.4 Non-trivial sets of disconnection forces corresponding to a mixed interface (left) and to a pseudo interface (right)

- Pseudo interface, including only internal DoFs ($i \subseteq r$) of subsystem R, e.g. those corresponding to the disconnection forces in Fig. 17.4 (right).

Note that the number of interface DoFs must be not less than the number of coupling DoFs n_c. The use of a mixed interface can be useful to replace rotational coupling DoFs with internal DoFs, while the use of a pseudo interface allows to replace all the coupling DoFs with translational internal DoFs.

Compatibility at the (standard, extended, mixed, pseudo) interface implies that any pair of matching DoFs, i.e. DoF l on the coupled system RU and DoF m on subsystem R, must have the same displacement, that is $u_l^{RU} - u_m^R = 0$. Let S be the set of N interface DoFs on which compatibility is enforced.

The compatibility condition can be generally expressed as

$$[B]\{u\} = \{0\} \qquad \text{where} \qquad \{u\} = \left\{ \begin{matrix} \{u^{RU}\} \\ *[1mm]\{u^R\} \\ *[1mm] \end{matrix} \right\} \tag{17.3}$$

where $[B]$ has size $N \times (N_{RU} + N_R)$ and each row corresponds to a pair of matching DoFs.

In the decoupling problem the equilibrium condition can be enforced on a different set of interface DoFs [7]; however in this case the same set S of N interface DoFs used for compatibility condition is considered. The equilibrium of disconnection forces implies that for any pair of matching DoFs, i.e. DoF r on the coupled system RU and DoF s on subsystem R, their sum must be zero that is $g_r^{RU} + g_s^R = 0$. Furthermore, for any unmatched DoF k on the coupled system RU (or on the residual subsystem R), it must be $g_k^{RU} = 0$ ($g_k^R = 0$). Altogether, the previous conditions can be expressed as

$$[L]^T\{g\} = \{0\} \qquad \text{where} \qquad \{g\} = \left\{ \begin{matrix} \{g^{RU}\} \\ *[1mm]\{g^R\} \\ *[1mm] \end{matrix} \right\} \tag{17.4}$$

where $[L]$ is a localization matrix and has size $(N_{RU} + N_R) \times (N_{RU} + N_R - N)$.

17.2.2 Formulation of the Decoupling Problem

The problem to be solved is mathematically obtained by gathering Eqs. (17.1-17.4). The solution is obtained using a dual assembly [1, 6] where equilibrium is satisfied exactly by defining a unique set of disconnection force intensities but compatibility could not be ensured.

In the dual assembly, the total set of DoFs is retained, and the equilibrium condition $g_r^{RU} + g_s^R = 0$ at a pair of matching DoFs is ensured by choosing $g_r^{RU} = -\lambda_r$ and $g_s^R = \lambda_r$. Therefore, the overall interface equilibrium can be ensured by writing the disconnection forces in the form:

$$\{g\} = -[B]^T\{\lambda\} \tag{17.5}$$

where $\{\lambda\}$ is a vector of Lagrange multipliers corresponding to disconnection force intensities.

Since there is a unique disconnection force intensity λ_r for any pair of equilibrium DoFs, Eq. (17.4) is satisfied automatically for any $\{\lambda\}$.

Therefore, Eqs. (17.1)–(17.3) become, after substituting Eq. (17.5) into Eqs. (17.1-17.2) recast in matrix form:

$$\begin{cases} [Z]\{u\} + [B]^T\{\lambda\} = \{f\} & (17.6) \\ [B]\{u\} = \{0\} & (17.7) \end{cases}$$

where $[Z]$ is obtained gathering the dynamic stiffness of the assembled system and that of the residual subsystem in block diagonal format as diag($[Z^{RU}]$, $-[Z^R]$), and $\{f\}$ is defined similarly to $\{u\}$ in Eq. (17.3).

To obtain the disconnection force intensity $\{\lambda\}$, displacements $\{u\}$ can be obtained from Eq. (17.6) and substituted into Eq. (17.7). Then, being $[Z]^{-1} = [H]$, $\{\lambda\}$ is found as

$$\{\lambda\} = \left([B][H][B]^T\right)^{-1}[B][H]\{f\} \tag{17.8}$$

The FRF of the unknown subsystem U can be obtained by back-substituting $\{\lambda\}$ in Eq. (17.6), and by isolating $\{u\}$ at the left hand side:

$$\{u\} = \left([H] - [H][B]^T\left([B][H][B]^T\right)^{-1}[B][H]\right)\{f\} \tag{17.9}$$

which is in the form $\{u\} = [H^U]\{f\}$, so that the FRF of the unknown subsystem U is

$$\left[H^U\right] = [H] - [H][B]^T\left([B][H][B]^T\right)^{-1}[B][H] \tag{17.10}$$

With the dual assembly, the rows and the columns of $[H^U]$ corresponding to compatibility and equilibrium DoFs appear twice. Furthermore, when using an extended or mixed interface, the rows and columns of $[H^U]$ corresponding to the internal DoFs of the residual substructure R are meaningless. Obviously, only meaningful and independent entries are considered. The matrix to be inverted is known as interface flexibility matrix, it depends on the choice of interface DoFs and it can be ill-conditioned for some set of interface DoFs.

17.2.3 Configuration-Dependent Decoupling

For a system that changes its configuration, it is possible to identify the dynamic of the same unknown subsystem using all the information provided by the different configurations. In order to apply a configuration dependent decoupling approach it is useful to consider a set of interface DoFs that is invariant as the configuration changes. Therefore, the selected interface DoFs should represent a pseudo interface for all the considered configurations, or a mixed interface for some configuration. Moreover, the DoF where the response of the unknown subsystem is observed should be an internal DoF of the unknown subsystem for all the considered configurations. With these assumptions, the matrix $[B]$ that enforces the compatibility conditions remains the same as the configuration changes. For each configuration χ, the FRF of the unknown subsystem U can be expressed as

$$\left[H^U\right] = [H(\chi)] - [H(\chi)][B]^T\left([B][H(\chi)][B]^T\right)^{-1}[B][H(\chi)] \tag{17.11}$$

where:

$$[H(\chi)] = \begin{bmatrix} [H^{RU}(\chi)] & [0] \\ [0] & -\left[H^R\right] \end{bmatrix} \tag{17.12}$$

i.e. only the FRF matrix of the assembled system $[H^{RU}(\chi)]$ is configuration dependent. In principle, $\left[H^U\right]$ is not configuration dependent; however, in practice some difference can arise, due to both numerical issues and noise in FRFs. By comparing the different solutions, it is possible to gain a better understanding of the dynamic behavior of the unknown subsystem.

17.3 Model and Configurations

In order to investigate the feasibility of configuration-dependent substructure decoupling a well-known test bed, previously used in [9] is considered. The assembled system consists of a cantilever beam with two short arms (the residual subsystem R) bolted to a beam (unknown subsystem U), shown in Fig. 17.5. The joint involves both translational and rotational DoFs. The cross section is 40 mm×8 mm for all beams, with the short side along the z-direction. The subsystem geometrical dimension is shown in Table 17.1.

Different configurations of the assembly can be obtained by connecting the residual and the unknown subsystems at different points. Figure 17.6 shows the three configurations of the assembly considered in the following of the paper and Table 17.1 shows the configuration dependent geometrical dimensions. Furthermore, Table 17.2 shows the coupling and the internal DoFs for the considered configurations.

The analysis is performed using simulated data, eventually polluted by random noise. FE models of the assembled structures, in different configurations, of the residual and of the unknown substructures are built using beam elements. The mechanical properties of the aluminum alloy are: $E = 7.0 \cdot 10^{10}$ N/m^2, $\rho = 2700$ kg/m^3, modal damping $\zeta = 0.005$.

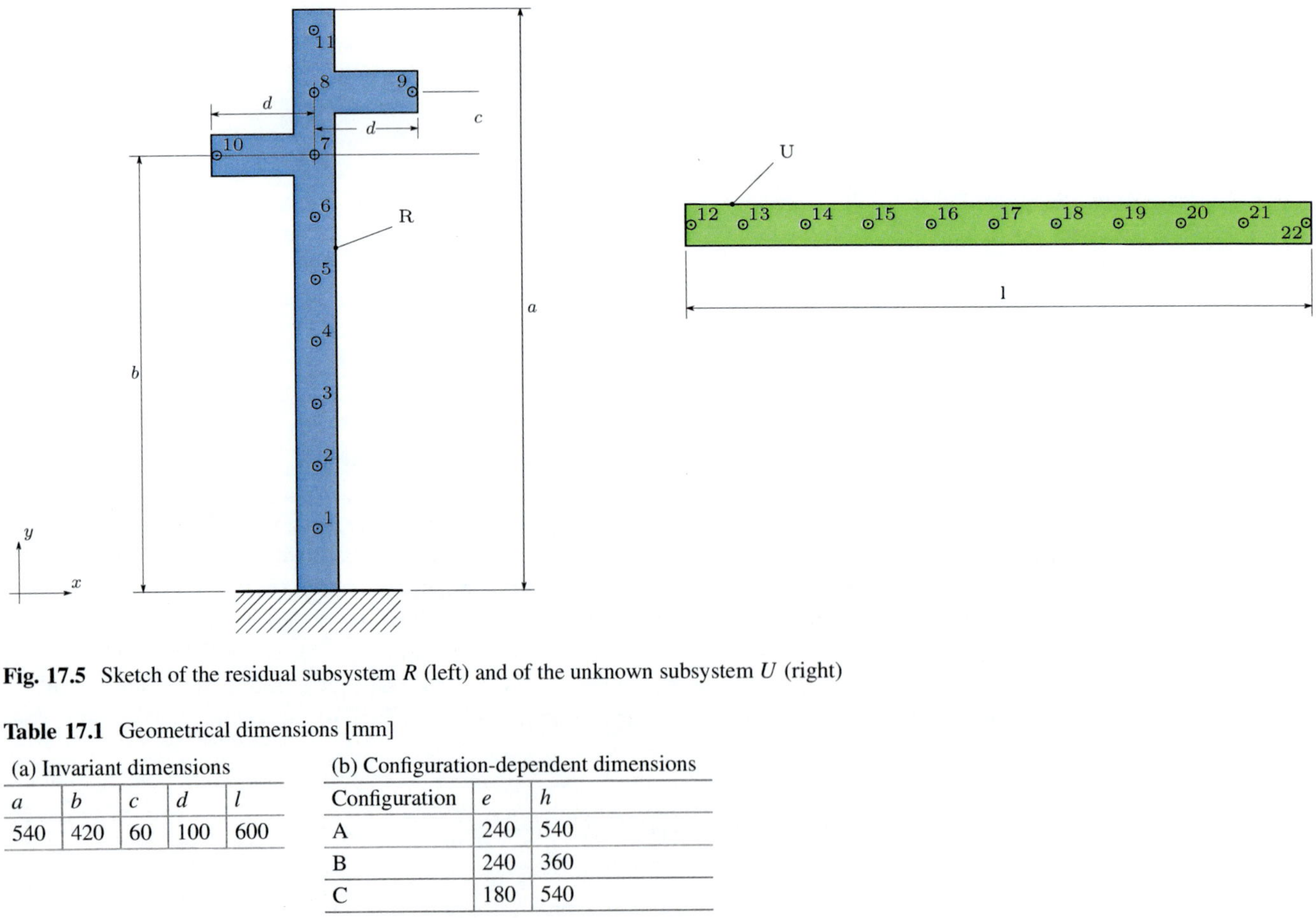

Fig. 17.5 Sketch of the residual subsystem R (left) and of the unknown subsystem U (right)

Table 17.1 Geometrical dimensions [mm]

(a) Invariant dimensions

a	b	c	d	l
540	420	60	100	600

(b) Configuration-dependent dimensions

Configuration	e	h
A	240	540
B	240	360
C	180	540

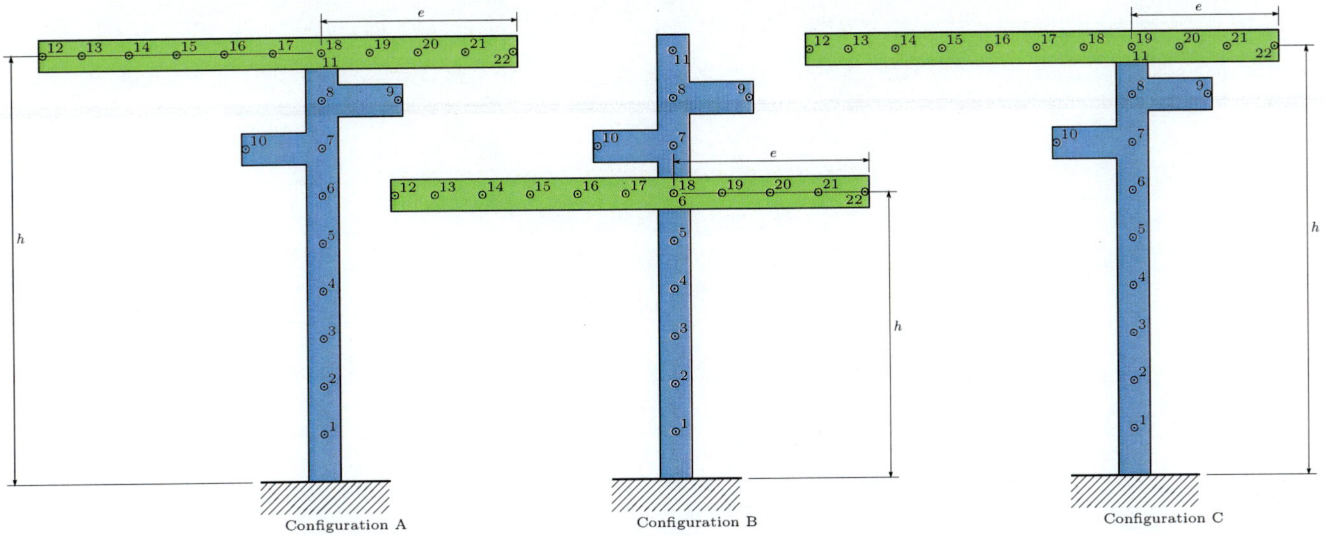

Fig. 17.6 Sketch of assembled system in different configurations

Table 17.2 Coupling and internal DoFs for the residual R and the unknown U subsystems for the three considered configurations of the assembly

Configuration	Coupling DoF R	Coupling DoF U	Internal DoFs R	Internal DoFs U
A	11	18	1:10	[12:17, 19:22]
B	6	18	[1:5, 7:11]	[12:17, 19:22]
C	11	19	1:10	[12:18, 20:22]

17.4 Results and Discussion

One of the most tricky aspect in the implementation of decoupling method is the choice of Degrees of Freedom to be used in the procedure. In fact, generally the coupling DoFs can be not directly measurable and they can involve also rotational DoFs. Therefore, it is necessary to identify a subset of translational internal DoFs of the residual subsystem R that should be used to substitute the unmeasured translational and rotational DoFs and in particular that of the junction point. Moreover, in this case different junction points are considered for the three configurations. The number n_c of coupling DoFs is 3 (translational DoF z and rotational DoFs ϑ_x and ϑ_y) so that the number internal DoFs to be considered to perform the decoupling must be $N_E \geq n_c$. In previous works it is shown that an analysis of FRF and Transmissibility criteria can be performed to select the optimal set of DoFs [17, 18]. In particular, the results obtained for the same structure considered in this paper in configuration A indicate that translational DoFs z of nodes 9 and/or 10 must be included in order to excite the rotational DoF ϑ_y at the junction.

For the three considered configurations the connecting points are 6 and 11, so that one of the valid set of internal DoFs of substructure R measured for decoupling is composed of DoFs $3z$, $7z$, $9z$, and $10z$.

Figure 17.7 shows the drive point inertance for DoF $16z$ of the unknown subsystem U evaluated considering the three different configurations. In general the decoupling procedure provides a good estimation of the main dynamics of the unknown subsystem U. In particular all configurations show slight differences between the phases of the true and the predicted FRFs over 700 Hz; in particular in configuration B there is a spurious peak in the magnitude of the predicted FRF at 733 Hz.

Since the decoupling procedure is very sensitive to the presence of noise on the measured data, the exact FRF of the assembled system and of the residual subsystem R, obtained using the numerical models, are polluted with 0.5% noise. Figure 17.8 shows the drive point inertance for DoF $16z$ of the unknown subsystem U evaluated considering the three different configurations using noise polluted FRFs. The results show that up to a frequency of about 500 Hz the main dynamics of the unknown subsystem is correctly estimated, but above 500 Hz different results are obtained with the three configurations. The analysis of results highlights that ill-conditioning of the decoupling problem, in presence of noise polluted data, amplifies the errors on the solution. For instance, the spurious peak observed in Fig. 17.7b for the configuration B using exact FRFs, in presence of noise becomes a wide band between around 600 and 900 Hz of unreliable results that

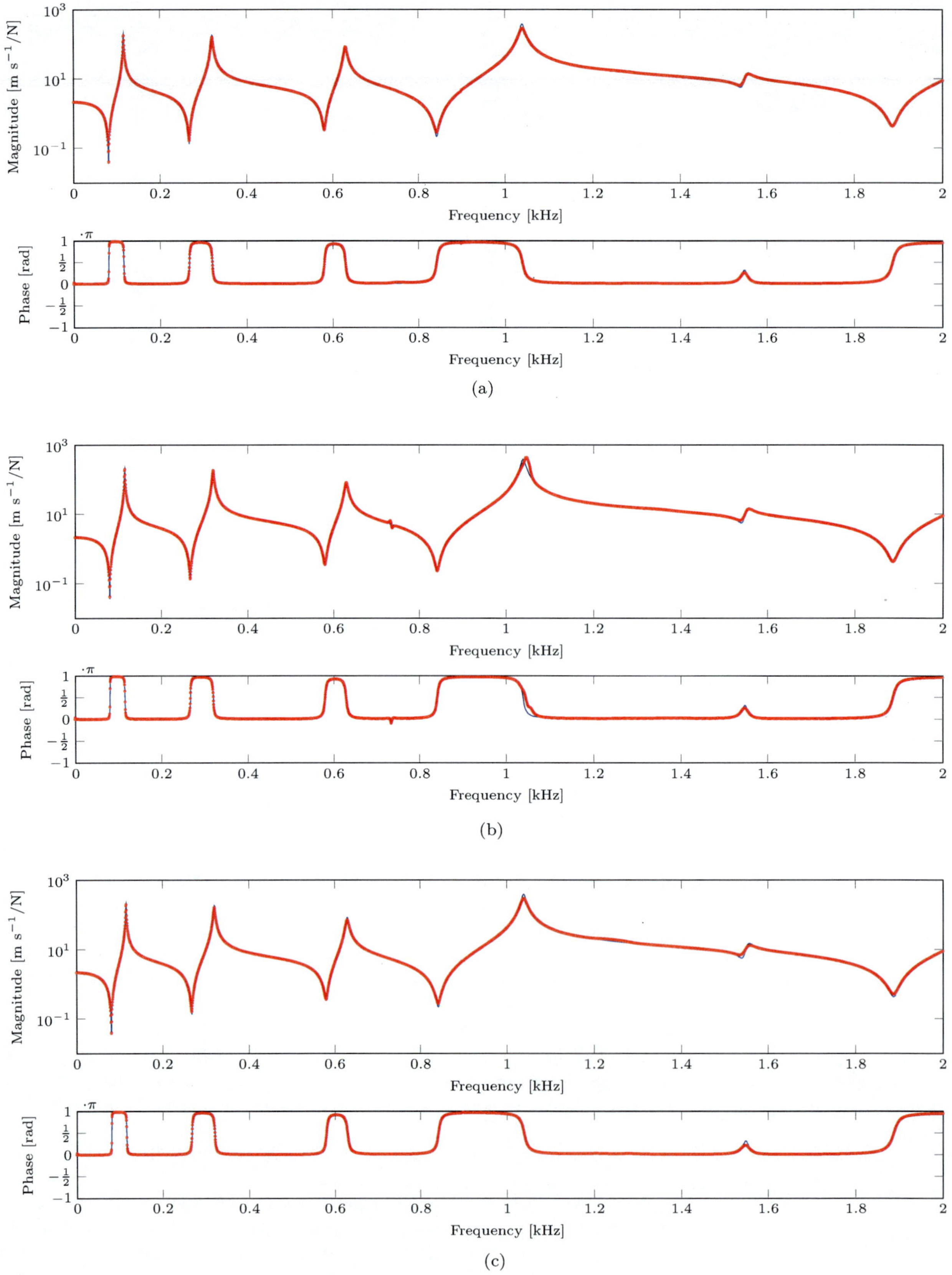

Fig. 17.7 Inertance $H^U_{16z,16z}$ of the unknown subsystem U: true solution (blue solid line), predicted from exact FRFs using internal DoFs $3z$, $7z$, $9z$ and $10z$ (red dots). (**a**) Configuration A. (**b**) Configuration B. (**c**) Configuration C

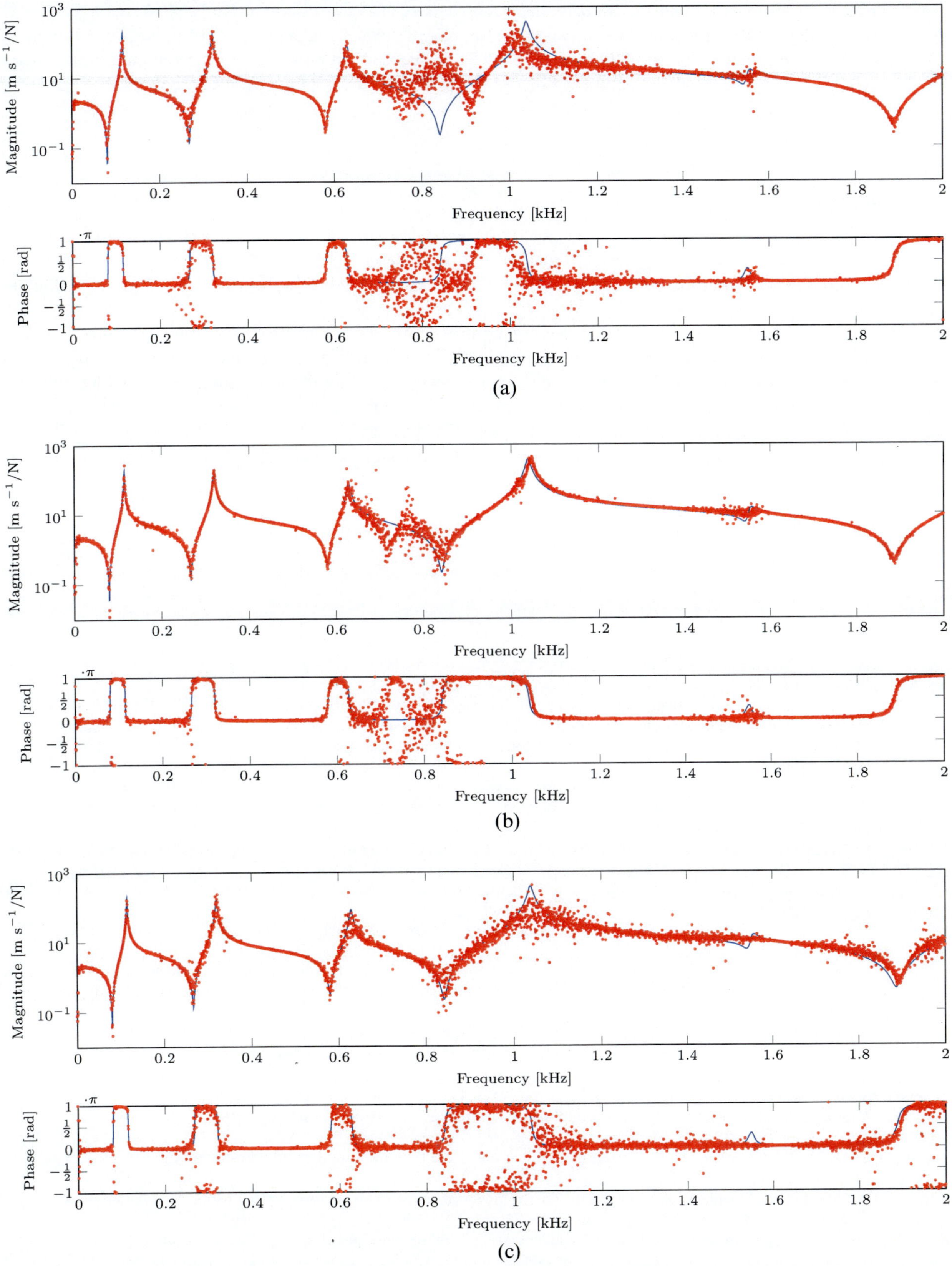

Fig. 17.8 Inertance $H^{U}_{16z,16z}$ of the unknown subsystem U: true solution (blue solid line), predicted from noise polluted FRFs using internal DoFs $3z$, $7z$, $9z$ and $10z$ (red dots). (**a**) Configuration A. (**b**) Configuration B. (**c**) Configuration C

completely shadows the dynamics around the third mode. Note that, even if the ill-conditioning produces a very slight perturbations in the predicted FRFs without noise, it can give rise to significant error in presence of noise. For example the slight phase distortion at frequencies 750 and 898 Hz in Fig. 17.7a, in presence of noise becomes even a fictitious mode at around 850 Hz. Among the three considered configurations of the assembly, the configuration C produces the best results (see Fig. 17.8c).

17.5 Conclusions

In this paper, the decoupling problem is applied to configuration dependent systems composed of invariant mechanical subsystems with configuration dependent coupling conditions. The decoupling procedure is applied to a system composed of two invariant substructures assembled in three different configurations. The decoupling procedure does not require measurements at the connecting coupling DoFs. The decoupling procedure can be performed to identify the same unknown subsystem, assembled in different configurations, without having to repeat measurements on the residual subsystem. Under these conditions the only difference introduced in the decoupling formulation by the configuration is the FRF matrix of the assembled structure. The comparison of the results obtained in different configurations of the assembled system can provide more insight on the dynamics of the unknown subsystem. Further developments will be devoted to the exploitation of the redundancy of information deriving from the multiple configurations of the assembled system, trying to use a global formulation.

Acknowledgments This research is supported by University of Rome La Sapienza and University of L'Aquila.

References

1. de Klerk, D., Rixen, D.J., Voormeeren, S.: General framework for dynamic substructuring: history, review, and classification of techniques. AIAA J. **46**(5), 1169–1181 (2008)
2. Craig, R.R.J., Bampton, M.: Coupling of substructures for dynamic analyses. AIAA J. **6**(7), 1313–1319 (1968). Cited By 2443
3. Rixen, D.: A dual Craig-Bampton method for dynamic substructuring. J. Comput. Appl. Math. **168**(1–2), 383–391 (2004). Cited By 156
4. Mayes, R.L., & Ross, M.R.: Advancements in hybrid dynamic models combining experimental and finite element substructures. Mech. Syst. Signal Process. **31**, 56– 66 (2012)
5. Sjövall, P., Abrahamsson, T.: Substructure system identification from coupled system test data. Mech. Syst. Signal Process. **22**(1), 15–33 (2008)
6. D'Ambrogio, W., Fregolent, A.: The role of interface DoFs in decoupling of substructures based on the dual domain decomposition. Mech. Syst. Signal Process. **24**(7): 2035–2048 (2010)
7. Voormeeren, S.N., Rixen, D.J.: A family of substructure decoupling techniques based on a dual assembly approach. Mech. Syst. Signal Process. **27**, 379–396 (2012)
8. D'Ambrogio, W., Fregolent, A.: Inverse dynamic substructuring using direct hybrid assembly in the frequency domain. Mech. Syst. Signal Process. **45**(2), 360–377 (2014)
9. D'Ambrogio, W., Fregolent, A.: Substructure decoupling without using rotational DoFs: fact or fiction? Mech. Syst. Signal Process. **72–73**, 499–512 (2016)
10. Brunetti, J., D'Ambrogio, W., Fregolent, A.: Contact problems in the framework of dynamic substructuring. In: Proceedings of ISMA 2018 - International Conference on Noise and Vibration Engineering and USD 2018 - International Conference on Uncertainty in Structural Dynamics, pp. 3987–3998 (2018)
11. Brunetti, J., D'Ambrogio, W., Fregolent, A.: Dynamic substructuring with a sliding contact interface. In: Dynamics of Coupled Structures, Linderholt, A., Allen, M.S., Mayes, R.L., Rixen, D. (eds.), vol. 4, pp. 105–116. Springer, Cham (2018)
12. Carassale, L., Silvestri, P., Lengu, R., Mazzaron, P.: Modeling rail-vehicle coupled dynamics by a time-varying substructuring scheme. In: Dynamic Substructures, Linderholt, A., Allen, M.S., Mayes, R.L., Rixen, D. (eds.), vol. 4, pp. 167–171. Springer, Cham (2020)
13. Brunetti, J., D'Ambrogio, W., Fregolent, A.: Dynamic coupling of substructures with sliding friction interfaces. Mech. Syst. Signal Process. **141**, 106731 (2020)
14. Brunetti, J., D'Ambrogio, W., Fregolent, A.: Friction-induced vibrations in the framework of dynamic substructuring. Nonlinear Dyn. **103**(4), 3301–3314 (2021)
15. Brunetti, J., D'Ambrogio, W., Fregolent, A.: Analysis of friction induced mode coupling instabilities using dynamic substructuring. In: Dynamic Substructures, Allen, M.S., D'Ambrogio, W., Roettgen, D. (eds.), pp. 91–97. Springer, Cham (2022)
16. Brunetti, J., D'Ambrogio, W., Fregolent, A.: Dynamic substructuring with time variant coupling conditions for the analysis of friction induced vibrations. In: Proceedings of ISMA 2020-International Conference on Noise and Vibration Engineering and USD 2020-International Conference on Uncertainty in Structural Dynamics, pp. 3023–3032 (2020)
17. D'Ambrogio, W., Fregolent, A.: Replacement of unobservable coupling DoFs in substructure decoupling. Mech. Syst. Signal Process. **95**, 380–396 (2017)
18. D'Ambrogio, W., Fregolent, A.: Substructure decoupling as the identification of a set of disconnection forces. Meccanica **52**(13), 3117–3129 (2017)